Eugène Dubois

The climates of the geological past and their relation to the

evolution of the sun

Eugène Dubois

The climates of the geological past and their relation to the evolution of the sun

ISBN/EAN: 9783744745437

Printed in Europe, USA, Canada, Australia, Japan

Cover: Foto ©berggeist007 / pixelio.de

More available books at **www.hansebooks.com**

THE CLIMATES

OF THE

GEOLOGICAL PAST

PREFACE.

The present essay is an attempt to explain by changes of the solar heat the great climatic changes of the Geological Past. It is a translation, with some small alterations, of a treatise in German: "Die Klimate der geologischen Vergangenheit und ihre Beziehung zur Entwickelungsgeschichte der Sonne", that appeared in the beginning of 1893 (Nijmegen, H. C. A. THIEME), — a new and enlarged edition of an essay in Dutch, which appeared in the beginning of 1891, under the title: "De klimaten der voorwereld en de geschiedenis der zon" in: "Natuurkundig Tijdschrift voor Nederlandsch-Indië". Deel 51, p. 37—92, at Batavia.

I take this opportunity of expressing my gratitude to my friend Mr. TH. DELPRAT for his kind assistance with the translation of the German text.

Both this text and the translation I have revised.

I hope this short essay to explain the great enigma of the climates of past ages may contribute towards its solution.

EUG. DUBOIS.

TOELOENG-AGOENG (Java),

CONTENTS.

PART I.

The Climates of Past Ages according to Palæontological and Geological Data.

The organisms of past ages as witnesses of climatic changes.—The fossil arctic florae as evidence of warmer climates.—Climatic zones of former periods.—Local anomalies in climatic conditions.—The hypothesis of a Tertiary displacement of the poles unnecessary and insufficient to explain the phenomena.—Gradual Tertiary cooling of the climates and Pleistocene Glacial epochs.

The Organisms of Past Ages as Witnesses of Climatic Changes.

IT has long been an established fact that the climates of past ages were different from the present. It is proved that in many regions of the globe during untold

millions of years a warmer climate prevailed
than that of the present time, and that
this afterwards gave place to a cooler
climate than the present. Nevertheless,
during some periods a general regularity
in the diffusion of heat over the earth,
resembling the present, is not to be mis-
taken.

We judge of the climates of past ages
chiefly by the fossil remains of plants and
animals. As a rule the organisms of warm
countries differ greatly from those of cold
regions, and, though not with indisputable
logic, we infer inversely that the climate of
a former period must have been warm or
cold according as its organisms are most
nearly related to those now existing in warm
or in cold regions. This inference is certainly quite justifiable when the relationship is
very close, and is apparent in a great many
species of plants and animals. In many

cases we may also infer considerable changes in the climate and its temperature from the greater or lesser energy which life in general displays. Now, we know for certain this general close relationship between fossil forms and organisms which exist at present in different climates, and also that many periods developed quite a different wealth of plant and animal life from the present flora and fauna.

Yet a fixed relation between the climate and the character of the organic world does not exist. Closely related plants and animals are met with in very different climates, and this was also the case in past ages. Elephants and rhinoceroses, at present true inhabitants of the tropical regions, existed in Europe during the Glacial epoch. Comparing parts of different continents situated in the same climatic zone with each other, we see that a nearly

perfect similarity of climate does not correspond at all with a similarity of organic
forms.   Further, plants and animals are,
in general, certainly most fitted for the
climate in which they exist, but it is
seldom known to what extent their existence
depends upon conditions which have no
causal relation with the climate.   Instances
are well known of European plants and
animals, which, under the most different
climatic conditions, have spread over the
whole globe, and have even caused the
extermination of indigenous species in their
new habitat.

Organic beings have, indeed, in a high
degree the aptitude for accommodating
themselves to the temperature of the climate.   In this respect the late eminent
geologist M. NEUMAYR, who contributed
much to the right appreciation of the data
according to which we judge the geological

climates, reminds us of a cause of error formerly pointed out by STARKIE GARDNER, DE SAPORTA, and others. Organic forms originating in a cooler climate may gradually spread southerly (in the northern hemisphere) whilst accommodating themselves to the new conditions. They acclimatise themselves to these new and on the whole easier conditions, and degenerate. In the meantime, new forms will have developed in their former home, which in their turn will pursue the same course and suffer the same fate. In this way southern types will always display a certain relationship to the older forms of the northern regions, the cause of which is thus not to be found in a change of temperature in their respective stations.

As early as 1856 HOOKER drew attention to the fact that in a uniformly warm and damp sea climate, luxuriant vegetable and animal life may exist at a lower mean temperature than in a dry continental climate. This accounts for the fact that the forests of New Zealand and Patagonia bear a subtropical character, the temperature yearly not being higher than that of France and England.*

From all this it will appear how cautiously we must proceed when attempting to draw any inference as to the temperature of a former age from the data of palæontology.† When used cautiously, however, these data are of great value, the greater the more recent the period is whence they

---

*Compare: M. NEUMAYR, Erdgeschichte. Vol. II. Leipzig 1887, and his lecture: "Die klimatischen Verhältnisse der Vorzeit." Schriften des Vereins zur Verbreitung naturwiss. Kenntnisse. Wien, 1889. Under the title of "The Climates of Past Ages" a translation of this lecture appeared in Nature, Vol. 42, 1890, pp. 148 and 175.

†See A. WOEIKOF, "Die Klimate der Erde." Vol. I. Chapter 12, Jena, 1887.

are obtained, as no important changes in the organisms can then have taken place. When we find in Central and Northern Europe, in a formation as recent as the Pleistocene, remains of the arctic willow *(Salix polaris)*, the dwarf birch *(Betula nana)*, of *Dryas octopetala, Polygonum viviparum,* and *Arctostaphylos uva-ursi,* together with the remains of true arctic mammals—the reindeer *(Cervus tarandus)*, the glutton *(Gulo borealis)*, the arctic fox *(Canis lagopus)*, the lemming *Myodes torquatus* and *obensis)*—and in the valleys the chamois and the ibex and also arctic and alpine snails; in the diluvial marine deposits such mollusks as *Pecten islandicus, Yoldia arctica, Astarte borealis* and *Saxicava arctica*—then we may safely conclude that in the Pleistocene period the climate of Europe was cooler than it is at present. When we furthermore find in the Middle Ter-

tiary formation in France and Switzerland—that is, in a comparatively recent
geological period—palms, fig-trees and
cinnamon-trees, together with other plants
with evergreen coriaceous leaves, tulip-
trees, *(Liriodendron)*, camphor-trees, laurels
and myrtles, like those which grow in our
days in countries with at least a subtropical climate, and together with many
other species—the remains of the gigantic
Dinotherium and *Helladotherium* of *Mastodon*
and *Rhinoceros*, together with anthropoid
apes; also trogons, parrots, pelicans, ibis-
birds and salanganes,—in the ancient sea
bottoms many species of mollusks of the
genera *Cypraea*, *Oliva*, *Mitra*, *Conus* and
Terebra, and near Vienna reef-building
corals of species still extant,—then from
these numerous facts we may safely conclude that also in this much more remote
period an entirely different, warmer climate

prevailed in Europe. Not taking into account the power of acclimatisation which species and genera undoubtedly possess, and the influence of the insular climate that prevailed in Europe then, we may, with HEER, conclude from the character of the Swiss Middle Tertiary flora, that the mean temperature there (reduced to sea level) must have been 7o to 9o C. higher than now. This power of acclimatisation, resulting from inconstancy of constitution and want of organisms, together with the favourable influence of a uniformly warm and damp climate, can scarcely account even partly for this difference. We must, therefore, take it for a fact that, in former periods, as well as in the tertiary period, the climates were warmer, not only in Europe, Asia, and North America, but also in the temperate zones of the southern hemisphere. Setting aside local irregularities in the distribution of heat, which existed then as now, the palæothermal phenomenon was a general occurrence which affected the whole of the globe, but so that the difference between the then and the present condition increased in the direction of the poles. This general greater equality of the climatic conditions in past periods is a settled fact.

The Fossil Arctic Floræ as evidence of Warmer Climates.

The best proof for this is given by the fossil floræ of the arctic regions, the description and interpretation of which has been one of the greatest achievements of geological science, and would alone have sufficed to immortalise the name of HEER.*

* Chiefly by his two principal works: " Die Urwelt der Schweiz," 1864 and 1879, and: " Flora Fossilis Arctica," 1868, OSWALD HEER must be considered as the founder

Gigantic Lepidodendra and Calamites, trees related to our wolfs' claws and mares' tails existed during the Carboniferous period, not only in the temperate zones of the

of Palæoclimatology. To him we owe nearly all the information which we possess about ancient climates. He proceeded with far greater caution in judging of the early distribution of heat over the earth from palæontological data than most of his successors, whose exaggerated conclusions are often posed as his.

Between 1871 and 1883 six volumes more of his "Flora Fossilis Arctica" were issued.

In a clear manner A. DE CANDOLLE, as early as 1855, in his "Géographie Botanique Raisonnée," and later in some smaller papers, thoroughly discussed the relation between the character of flora and climate, and the ideas of this great botanist of Geneva exercised a strong influence over HEER'S labours.

The first indications of a previous milder climate in higher latitudes were found by A. ERMAN, in 1829, in Kamtchatka at 63° N.L. ("Reise um die Erde," Vol. 3. Berlin, 1848, p. 149).

A small treatise of H. R. GOEPPERT ("Ueber die Tertiärflora der Polargegenden," in : Bulletin de l'Académie Imp. des Sciences de St. Pétersbourg. T. 3. 1861, p. 448—461) had remained almost unnoticed.

northern and southern hemispheres, but also at Spitzbergen at $77\frac{1}{2}^0$ N.L. Of the 26 species of Carboniferous plants which were collected there, 17 existed during the same period in Central Europe. Further, a luxuriant Carboniferous vegetation existed within the American north polar circle at 75^0 latitude, and in Siberia near the mouth of the Lena at $71\frac{1}{2}^0$ N.L.

Fossil plants from the Permian and Triassic period have not as yet been met with in arctic regions, but two species of *Ichthyosaurus* were found at Spitzbergen at $78\frac{1}{2}^0$ N.L.; and marine mollusks, equal to those which existed during that period in temperate zones, were found chiefly in North Siberia near the mouth of the Olenek at 73^0 N.L., also at Spitzbergen.

In the Jurassic formation of Spitzbergen 32 species of mares' tails, ferns, cycads, and conifers were found at $78\frac{1}{2}^0$ N.L.,

10 of which were also met with in temperate latitudes.

Of more importance, however, is our knowledge of the arctic Cretaceous formation. HEER made us acquainted with the rich Cretaceous flora, which NORDENSKIÖLD collected on the west coast of Greenland at 70⁰ N.L., consisting of ferns, mares' tails, cycads, conifers, grasses, rushes, together with one Dicotyledonous plant *(Populus primaeva);* and, from somewhat more recent beds, many Dicotyledonous trees, such as magnoliæ, different species of fig-trees, sassafras and poplars, similar to those found in the later Cretaceous formation of Europe, where, however, tropical and subtropical species are more numerously represented.

Still richer is the Tertiary vegetation of the arctic regions, which HEER inferred to be Miocene, but to which some pa-

læontologists, with BELT, STARKIE GARDNER
and DE SAPORTA, now ascribe a somewhat
older, *viz.*, Eocene, character.   HEER described Tertiary plants from Iceland, Greenland, Grinnell Land, Spitzbergen, and North
Canada.   Grinnell Land, in the great North
American archipelago, is the most northern
point of the globe, from which a fossil
flora is known.   At $81\,^3/_4{}^0$ N.L. Captain
FEILDEN, member of the NARES Arctic
Expedition, collected there 26 species of
plants, which were afterwards described by
HEER.   Ten were found to be conifers,
the gigantic swamp-cypress *(Taxodium
distichum)*, which grows at present in the
Southern United States, forming a large
portion of them.   Together with these
grew poplars, birches, limes, elms, hazels
and a Viburnum, and in a lake enclosed
by the same forest, water-lilies, sedges
and rushes.   A similar vegetation pre-

vails at present in the northern temperate zone with a mean yearly temperature
of 8⁰ C; yet such a vegetation formerly
existed in a country where the mean temperature is at present 20⁰—C!

HEER described 179 species of Tertiary
plants from Spitzbergen, from $77\frac{1}{2}^0$ to 79^0
N.L., which belong to a flora almost similar to
that of Grinnell Land but pointing to a
slightly warmer climate. True cypresses
(Libocedrus), willows, and not less than
seven species of poplars, together with
oaks, planes, magnoliæ, walnut-trees and
maples; also cornelian, hawthorn and other
shrubs were found, and in a swamp, reeds,
orris roots, calamus and *Potamogeton*. This
indicates a flora similar to that which prevails at present in Northern Germany with
a mean yearly temperature of 9⁰ C. The
mean yearly temperature in that part of
Spitzbergen is now—8⁰ C.

The rich Tertiary flora of Greenland (especially at 70⁰ N.L.) must have grown in an even somewhat warmer climate. The present mean temperature there is — 7⁰ C. There grew, side by side with many other species of large forest trees, the mammoth-tree *(Sequoia)*, the Japanese genus Gingko *(Salisburea)*, chestnut-trees, magnoliæ, evergreen oaks, laurels and a vine. HEER points out that nowadays we must travel 20 to 25 degrees of latitude southwards, to the shores of the lake of Geneva, to California and Japan, to meet in Europe, North America and Asia, at a mean yearly temperature of $10\frac{1}{2}^0$ C., a similar vegetation.

Further west in the American arctic archipelago, at Banks Land (74⁰ N.L.) and at Prince Patrick Island (76⁰ N.L.)— and further east at the Island of New Siberia (at 75⁰ and $75\frac{1}{2}^0$ N.L.), and at some other places within the Arctic circle, numerous remains of a rich Tertiary forest vegetation have been collected, from which we may conclude that everywhere around and near the north pole a much warmer climate than the present must have prevailed, and that the difference from the actual condition was much more considerable there than we find it to have been in temperate latitudes.* Whilst a rise in temperature of 7° C. would have sufficed to account for the Middle Tertiary vege-

*As far as Banks Land and Prince Patrick Island are concerned some possibility remains that the trees found had been floated there, though the discoverers themselves, MAC-CLURE and MACCLINTOCK, were of opinion that they had grown on the place where they were found. According to VON TOLL this must certainly have been the case in New Siberia, for together with those trees brown coal and beds consisting merely of the remains of leaves and fruit-cones occur. (J. SCHMALHAUSEN, "Tertiäre Pflanzen der Insel Neu-Sibirien," Mémoires de l'Acad. Impér. des Sc. de St. Pétersbourg. T. 37. 1890, No. 5.)

tation in Switzerland, it would have been far from sufficient for the Arctic regions. The difference from the present condition was so great there, that we must assume the temperature of Greenland, at 70° N.L., and of Spitzbergen, at $78\frac{1}{2}^\circ$ N.L., to have been at least 17° C., and that of Grinnell Land, at $81\frac{3}{4}^\circ$ N.L., even 28° C. higher than the actual temperature. In considering these figures we must keep in mind that the climate of North America at present is relatively cold compared to that of Europe, and that, if the North American archipelago were now favoured with warm ocean currents like Spitzbergen and the coast of Greenland, a rise of temperature for Grinnell Land equal to that of the more easterly Arctic regions would have sufficed.

We observe that, as a general rule, the difference between the past and the actual temperature of any climate increases, the nearer we approach the pole. Judging by the character of Middle Tertiary plants from Java, Borneo and Sumatra, described by GOEPPERT, HEER, GEYLER, CRIÉ and VON ETTINGSHAUSEN, and of Miocene sea animals from Java, described by MARTIN, the difference between the temperature then prevailing near the equator with the present can have been but trifling. According to certain observations made with regard to the distribution of organisms, especially land organisms in the Siluric, the first half of the Coal period, and some earlier periods, this difference may have been a little greater inland during still earlier periods. *

Very little is as yet known of early life

* According to later investigations by HEER (Compare: C. SCHRÖTER, "OSWALD HEER, Lebensbild eines schweizeri- schen Naturforschers." Zürich, 1888, p. 335), the temperatures in the Tertiary period (Lower Miocene and Oligocene of HEER) were:

in high southern latitudes; but judging by
the character of the fossil remains of plants
and animals found in temperate latitudes,
conditions in general similar to those
observed in the northern hemisphere must
have prevailed in Antarctic regions. *

Northern Italy . . . . 22° C.

Switzerland . . . . 20$^{1}/_{2}$°

Basin of the Lower Rhine . . . . 18°

Dantzig . . . . 17°

Greenland at 70° N.L. . . . . . 12°

Spitzbergen at 78° N.L. . . . . . 9°

Grinnell Land at 82° N.L. . . . . 8°

*The Tertiary flora of New South Wales and of New
Zealand had, for instance, the same climatic character as that
of Europe, and was even closely related to it. (VON
ETTINGSHAUSEN, Denkschr. K. Akad. Wien, Math. Nat.
Cl. 1886. Vol. 53, p. 79, and Ibid. 1887, p. 140.—From
the Tertiary formation in Kerguelen Island L. (49° S.L.) CRIÉ
(Palæont. Abhandl. von DAMES und KAYSER, N. F. Vol. 1.
Heft 2. 1889) described large stems of trees (*Cupressoxylon*),
and from Punta Arenas (53$^{1}/_{2}$° S.L.), at the Strait of
Magellan, H. ENGELHARDT (Abhandl. der Senckenberg.
naturf. Gesellsch. Frankfurt a. M. 1891) Vol. 16, pp. 629—
672 a Tertiary florula of tropical character.

An attempt to explain that even ever-
greens and shrubs could thrive in
Arctic regions, notwithstanding the long
polar night, is made by HEER from a
comparison with present facts; namely,
that even now some trees and shrubs
survive the long polar night near the 70th
degree of northern latitude; that many
tropical plants live during the long winter in St. Petersburg in hot-houses, while
the amount of light received is but very
small; and that evergreen alpine shrubs,
mountain-pine *(Pinus pumilis),* rhododen-
drons and heather are likewise for many
months cut off from all light by a covering
of snow. To these arguments we may
add that the polar night is considerably
shortened by a very strong twilight, *
that the lighting power of solar rays at

* Compare: J. HANN, "Handbuch der Klimatologie."
Stuttgart, 1883, p. 749.

a low angle of incidence decreases considerably less than the heating power, and that lastly, and perhaps chiefly, though the arctic summer be of short duration, the quantity of solar light received is comparatively large, and equal to half the quantity received in the same length of time near the equator.* The arctic plants are thus enabled to store a large supply of fecula, by which they can bring buds and leaves to rapid development, as soon as the temperature rises in the spring.†

Near the pole the solar radiation in midsummer is daily greater than it ever is at the equator. The greater part of this radiant heat is now employed in melting

*R. SPITALER, "Die Lichtvertheilung auf der Erdoberfläche," in: EDER's Jahrbuch für Photographie, 1888.

† Compare the picture of the relatively luxuriant present vegetation in the interior of Grinnell Land, drawn by A. W. GREELY in: "Three Years of Arctic Service." London, 1886.

the ice and the large quantities of snow which have been formed and have accumulated during the long frosty seasons. How entirely different would vegetation flourish in polar regions if during the whole year warm sea currents prevented the formation of the huge ice masses, and mild winds continued blowing during the long winter night!*

Climatic Zones of former Periods.

One of the most important results of HEER's investigations is that the Tertiary arctic flora forms a circuit around the pole, which fact caused that great phyto-palæontologist to reject emphatically any displacement of the earth's axis of rotation and geographical displacement of the poles, suggested as an explanation of the Tertiary distribution of heat.

* Compare: HANN, "Handbuch der Klimatologie," p. 744.

He also proved the existence of climatic zones concentric around the present pole during the Tertiary and the last half of the Cretaceous periods. He was of opinion —an opinion that is now still too widely spread—that in the former periods over the whole surface of the earth a uniform tropical climate prevailed, though with a range of the mean yearly temperature from 20^0 to 25^0 C. The groundlessness of the assumption of a perfectly uniform distribution of heat during the Pretertiary ages has since been recognised. Besides irregular climatic anomalies in earlier periods, a distinct northern and southern facies of the fauna has been observed to have existed during the Cretaceous period (F. RÖMER 1852), which must be ascribed to a similar, though less pronounced distribution of heat in zones between the equator and the poles as now exists.

NEUMAYR,* further, amply proved the existence of heat belts during the Jurassic period. By more special investigation of the distribution of marine animals he succeeded in recognising a tropical zone, which reached southward as far as the 20th degree of southern latitude and included northward the southern part of Europe; a temperate (as regards climate: subtropical) zone, which reached farthest north in a large sea arm to the west of Scandinavia, corresponding to the present Atlantic Ocean; and a boreal (as regards climate: warm temperate) zone which included the highest northern latitudes.

During the entire Jurassic period which, judging by the changes observed in its

---

* M. NEUMAYR, " Ueber klimatische Zonen während der Jura- und Kreide-Zeit." Denkschriften der K. Akademie der Wissensch. zu Wien. Math. Nat. Cl. 1883. Vol. 47, and: Die geographische Verbreitung der Juraformation. Ibid. 1885, Vol. 50, p. 57.

organisms, must have lasted much longer than the Tertiary period, and also during the Cretaceous period, these limits remain on the whole unaltered, notwithstanding the considerable changes which took place in the distribution of land and sea. This proves how little influence distribution exercises on the general diffusion of heat over the surface of the earth, at least as far as the temperature of the sea water (and sea climate) is concerned.

Probably during earlier periods similar zones, with presumably still less mutual differences in temperature, existed—and it is quite possible that some day we may succeed, in other parts of the world, in proving their existence, in the same way as that of the zones during the Jurassic and the Cretaceous periods in Europe. As has been rightly said by NEUMAYR the geological circumstances are too unfavourable in Europe for arriving at an estimation of the climates in earlier periods. It would really seem as if the Coal flora and the fauna of the Carboniferous limestone in high northern latitudes differed slightly from those of Europe, by the scarcity or absence of *Sigillariæ* and the scarcity of *Fusulinæ*. On account of the almost total stability observed during the greater part of the Mesozoic period we may assume as probable that the distribution of heat since the beginning of the Palæozoic period also has undergone but slight changes. So much is now manifest, that a lower temperature in Europe and North America during the Coal period than the present temperature in the tropical zone would best explain the observed phenomena; but it would be too much to infer that a New Zealand forest is the best recent illustration of the old Coal vegetation. Not the three ferns and conifers preponderating there, but trees, having succulent stems, whose near relations, the present herbaceous *Equi-setaceæ* and *Lycopodinæ*, are mostly inhabi-tants of the tropics, we must consider as the chief types of the Carboniferous forma-tion. We might somewhat more rightly compare the Coal flora to the swamp forests of gigantic *Equisetaceæ* near Caracas in Venezuela. The Carboniferous ferns differ considerably from the present types, but half the number of their species were related to the recent tropical family, the *Marattiaceæ*. Every closer comparison of that early flora with plants now existing must necessarily fail, through their being too distantly related, but we may safely conclude from the general character of the Carboniferous flora that those old crypto-gamiæ with their gigantic but succulent stems, with little wood fibre, sparingly

branched and having short, stiff, grass-like
leaves, may have been fit to resist a strong
wind but certainly not even the slightest
frost.   The fact that these existed in temperate latitudes and even in the polar region
seems to prove conclusively that there,
during the Palæozoic period, not only a
more moist but also a considerably warmer
and more uniform climate than the present
must have prevailed.   A proof, almost
equal to an experiment, for the correctness
of the assumption that a mild climate has
been actually a vital necessity for the Coal
flora, is given by what took place at the
incidence of glacial phenomena in South
Africa and Australia during the last part
of the Carboniferous period.   With the
approaching cold we see the real Palæozoic
Coal flora, which at that time was everywhere else at its highest state of development,
immediately and entirely disappear there

and give place to a flora which appeared in the Triassic period* in Europe much later on.

Not only the distribution of plants, generally more directly dependent upon the climate for existence than is the case with animals, but also the distribution of reef-building corals has long been rightly considered an important phenomenon by which to judge of early climates; corals being characteristic tropical organisms. Reef-

---

*W. WAAGEN, "The Carboniferous Glacial Period. Rec. Geol. Surv. Ind." Vol. 21. 1888, p. 89.—Compare also: O. FEISTMANTEL. "Ueber die pflanzen- und kohlenführenden Schichten in Indien, Afrika und Australien und darin vorkommende glaziale Erscheinungen." Sitz. Ber. Böhm. Gesellsch. d. Wiss. 1887, p. 3, and: "Ueber die geologischen und palæontologischen Verhältnisse der im Gondwána-System vorkommenden Arten." Sitz. Ber. Böhm. Ges. d. Wiss. 1889, p. 584—654, further: "On the Coal and Plant-bearing Beds of Palæozoic and Mesozoic Ages in Eastern Australia and Tasmania; with special reference to the fossil Flora. Mem. Geol. Surv. New South Wales." 1890.

building coral zoophytes live at present only in warm seas, the surface temperature of which never falls below 20° C., and where the annual range of temperature is not greater than 7° C. This accounts for the fact that they are found as far as the Bermuda Islands at 32° N.L. where the influence of the warm Gulf Stream is still very noticeable, and not near the Galapagos Islands under the equator, the coasts of which are washed by the cool Peru current. Enormous old coral reefs are known from the Silurian and Carboniferous periods of Europe, North America and many other regions in temperate and high latitudes. Within the Arctic circle Silurian reef-building corals have been found at North Devon and Beechy Island. But the type of these Palæozoic corals differs very essentially from those now existing; for all this, it might be possible that they had accommodated themselves to conditions different from those under which the present reef-builders exist, though on physiological grounds we can scarcely think that this difference can have been very considerable.*

Remains of old coral reefs from the Triassic period exist in Central Europe.

The temperature of early seas may be concluded with greater certainty from the distribution of coral reefs in the Jurassic and Tertiary formation, as these belong to the present type. Those of the Jurassic period reach as far north as Northern Germany and Central England. In the Cretaceous formation and in the first half of the Tertiary period their limit was but

*Because the secretion of lime by the organisms is in a high degree dependent on the temperature. See: MURRAY and R. IRVINE, "On Coral Reefs and other Carbonate of Lime Formations in Modern Seas." Proceed. Royal Society of Edinburgh. Vol. 17. 1890, pp. 80—82 and p. 90.

a little more southerly, while on the contrary in other parts of the world their limit was farther from the equator. General changes (having affected the whole of the globe) did not occur—as far as we know—during the Cretaceous period.

Local Anomalies in Climatic Conditions.

The climate of North America was in the Jurassic period already somewhat cooler than that of Europe, a difference which became still more apparent in the Cretaceous period, and while an almost tropical climate prevailed in Central Europe during the Eocene period, the Eocene flora of the North American continent bore a subtropical or moderately warm character. During the Tertiary period, at least, the climate of Europe compared to that of other regions must have been abnormally warm and that of the interior of North

America abnormally cool—a difference
entirely analogous to that observed nowadays, which in all probability is attri-
butable to the same causes. Already
in very early times North America formed
an extensive continent, which developed to
its present condition during the transition
of the Cretaceous to the Tertiary period
when the last extensive bays receded. It
is therefore obvious that, under the same
general conditions of heat on the earth as
prevail at present, its climate must have
been quite different from—very much cooler
than that of Europe,—for this consisted
for long ages of islands and peninsulas
with extensive inland seas and large bays
into which warm water was driven from
southern regions by the prevailing winds,
rendering the climate a very mild one
even as far as the polar sea. It seems
probable, moreover, that during long ages,

and as late as the Tertiary age, extensive connections existed through Russia and Siberia between the arctic seas and the warm seas of Europe and Southern Asia. Hence, even for the arctic regions of North East Asia and North America more favourable conditions must have prevailed than at present. This explains the fact that in Central Asia at the foot of the Altai mountains a true Tertiary forest vegetation could exist.*

Local anomalies in the general distribution of heat obtained in every former period the same as they do now, and so many geologists have found evidences of ice action in different parts of the globe and so nearly in the complete series of formations, that these cannot well be passed over unobserved.† Those earlier ice forma-

*SCHMALHAUSEN and MAXIMOWICZ in; Palæontographica. Beiträge zur Naturgeschichte der Vorzeit, herausgeg. von K. A. VON ZITTEL. Vol 33. Stuttgart 1885, pp. 142—216.

† A summary of those cited pre-Diluvial glaciations may

tions, however, differ in so far from the
Pleistocene glaciation, that they do not
occur as a general phenomenon, but scattered over various parts of the globe,
apparently independent of any special latitude, while at the same time organisms
indicative of a warm climate existed in
other regions under very different latitudes.
From the last half of the Carboniferous
period (according to some, the first Permian period) extensive and thick beds
exist in South Africa, India and Australia,
which do not contain Sigillariæ, Lepidodendra, Calamites, or other typical coal
plants, but in which is found a flora of
ferns, chiefly belonging to the genus
Glossopteris, with its allies *Gangamopteris*
and *Nöggerathiopsis*, together with Equisetaceæ, some Conifers and Cycads. This

be found in: CROLL, "Climate and Time." London, 1875.
Chapter XVIII.

vegetation, as above mentioned, did not at
all belong to the Carboniferous flora, but is
nearly related to that which was to appear
later on in Europe during the Triassic
period. This new flora is found in South
Africa, India and Australia together with
deposits which to all appearance have been
formed by ice action.* During the same

---

* See: W. WAAGEN, "The Carboniferous Glacial Period."
l.c.; O. FEISTMANTEL, "Ueber die pflanzen-und kohlenführenden
Schichten in Indien, Afrika und Australien, l.c., etc."

According to later communications traces of a Carboniferous glaciation were supposed to have been found in Brazil.
"(The Carboniferous Glacial Period." Further Note by Dr.
W. WAAGEN on a letter from Mr. A. DERBY. "Records
of the Geol. Survey of India." Vol. 22. 1889, p. 69). The
presence of large blocks of granite and gneiss in fine clay
can, however, be explained by other causes than ice-trans-
port. As long as no clearly polished rocks and scratched
stones have been found it will be prudent to consider the
occurrence of glaciation during the Carboniferous formation
of Brazil (though this contains very considerable beds of
Coal) as at least very doubtful. For pseudo-glacial phenomena see: A. PENCK in: "Das Ausland" 1884, p. 641 sqq.

time the luxuriant Carboniferous vegetation
existed in Europe, Asia and North America,
and even close to the north pole, and the
Carboniferous marine fauna remains unaltered. This remarkable phenomenon can,
therefore, not be explained by one general
cause, the influence of which was felt over
the whole earth. Here again a displacement of the earth's axis and the poles has
been assumed.* On closely investigating
the effect of such a displacement it becomes apparent that, though for one of the
poles a geographical position might be
suggested by which the simultaneous existence of ice and a nearly tropical vegetation on different parts of the earth, where
quite opposite circumstances obtain at
present, could be made somewhat less
obscure, yet this suggestion would neces-

*H. B. MEDLICOTT and W. T. BLANFORD, "Manual of
the Geology of India." Part. I. Calcutta, 1879, p. xxxvii.

sarily put the opposite pole in the middle of a typical Coal flora. Further, the distance of the ice-covered regions from the hypothetical south pole is so great that the temperature of the whole earth ought to have been much lower than it is at present, which certainly was not the case. It seems quite impossible to explain these phenomena by general causes, and we are obliged to ascribe even those extensive climatic anomalies to local influences. Now, it has been proved that the existence of ice in the form of glaciers depends much more upon the dampness of the climate than upon its low temperature. In the moist regions of New Zealand, with a mean yearly temperature of 10° C.—a climate equally as warm as that of Vienna and warmer than that of Geneva—glaciers descend as low as 213 metres above the sea level, in between dense forests with fern,

trees, palms and fuchsias, while in the dry
mountain ranges of Eastern Siberia, with
a mean yearly temperature varying from
— 15⁰ C. to — 18⁰ C., neither permanent snow
nor glaciers are found.   On the southern
slopes of the Himalayas (with a mean
yearly temperature of +0.5⁰ C. at snow
line) permanent snow and glaciers descend
more than 700 metres lower than on the
northern slopes (with a temperature of
— 2.8⁰ C. at snow line).   The yearly temperature near the limit of permanent snow
is in the tropics +1.5⁰ C. and in the
Swiss Alps about — 3⁰ C.   It has also been
observed that the periods during which
the glaciers in the Alps increase may coincide with warm years, and that on the
contrary periods during which glaciers
decrease may correspond with cold years;
but an increase of glaciers takes place
when during some years moist westerly

winds have prevailed, and a decrease when easterly and northerly winds have been prevalent.* These and other phenomena justify the conclusion that the main factor in the formation of glaciers is the amount of snow fallen in the fern region. If in early periods, and as late as in the Tertiary period, heat was distributed far more uniformly over the earth, without any decrease of temperature being observable in the equatorial regions, tellural causes being, further, totally inefficient in themselves to affect any considerable change in the general temperature of the globe, then it is obvious that the total amount of heat received by the earth must have been greater. The atmosphere must then have contained a greater amount of vapour,

* C. LANG, "Der säkuläre Verlauf der Witterung als Ursache der Gletscherchwankungen in den Alpen." Zeitschr. der Oesterr. Gesellsch. für Meteorologie. 1885. Vol. 20, p, 443.

and in places where a considerable upheaval
of the surface of the earth existed snow
and ice could accumulate to form glaciers
more easily than at present.   In regions
which were as warm then as the tropics
now are the lower limit of permanent snow
may have been below the present (5000
metres) and the upper limit, on the contrary, above it.   With an abundant supply
of snow it may then have been possible
that in mountainous regions glaciers were
formed, which under very favourable circumstances descended even as low as the
sea level.   Just as during the Carboniferous age an extensive lowland, cut up by
the sea into a large marshy archipelago,
accounts for the formation of coal over
nearly the whole of the northern hemisphere, to such an extent that comparison
can only be made with the extensive deposits of Jurassic coal, extending from Western

Asia to Australia, it seems that a large
mountainous continent (" Gondwána Land "
of SUESS), at and south of the equator, has
caused extensive accumulations of ice in
suitable places. A great uniformity of
orographic conditions over extensive continental parts of the earth's crust seems to
have been characteristic of the Coal period.
It is thus possible, and even probable, that
by a gradual upheaval of such a continent,
the changed conditions of existence caused
the development of a new flora, which
only much later, in the beginning of the
Mesozoic period, should find in Europe,
in the higher upheaval of the ground, conditions it was better fitted for than was
the older Palæozoic flora which in consequence would suffer extermination.   Traces
of glaciation are believed to have been
actually found in the Permian formation
of Europe.   From those high centres of

acclimatisation the new flora, accommodat
ing itself to a higher temperature, could
then have gradually spread over the low-
lands.

Many particulars in the distribution o
organisms of other periods also point to
differences of temperature in regions not
far distant from one another, and as no
difference in latitude can account for these
anomalies, they must be attributed to local
influences similar to those caused in our
days by the distribution of land and sea,
the elevation of the land, the direction of
winds, and the neighbourhood of cold or
warm ocean currents. As far, however,
as we were able to detect general regularity in the distribution of heat, with
certainty, since the Jurassic period,
the distribution then was similar to that
now prevailing, and formed concentric
belts round the poles, differing only from

the present by their mutual limits and differences of temperature.

The Hypothesis of a Tertiary Displacement of the Pole unnecessary, and insufficient to explain the Phenomena.

It was already known to HEER that the Tertiary arctic flora formed a chain round the pole from which it could nowhere escape. Neither would the supposition of a displacement of the pole account here for the palæothermal phenomenon;* but,

---

*An acccount of the earlier attempts to explain the warm arctic climates by the secular changes in the position of the earth's axis has been given by F. VON CZERNY ("Die Veränderlichkeit des Klimas und ihre Ursachen." Wien. Pest. Leipzig 1881, p. 86 sqq.). S. HAUGHTON. as early as 1878 ("Physical Geography, Nature." Vol. 18, p. 266 sqq.) proved on geological grounds the impossibility of this explanation.— With regard to the mechanical side of this question it is now certain that any important deviations of the poles can only be caused by geological processes if the rigidity of the earth is not very considerable. See: G. V. SCHIAPARELLI.

as this could be explained by no other hypothesis, the almost-forgotten displacement of the pole was again in recent times seized upon, just as a drowning man grasps at a straw. Encouraged by new astronomical data, by which very small displacements of the earth's axis were proved, NEUMAYR and NATHORST considered it warrantable to attempt an explanation of the polar regions, at least partly, by similar, but larger, displacements.* It is, however, not only very questionable whether one is justified in assuming a lasting displacement of ten and more degrees,

---

"De la rotation de la Terre sous l'influence des actions géologiques. Mémoire présenté à l'Observatoire de Poulkova à l'occasion de sa fête semiséculaire. St. Pétersbourg 1889.

*M. NEUMAYR, Erdgeschichte. Vol. II, p. 512 sqq. and: "The Climates of Past Ages. Nature." Vol. 42. 1890, p. 171.

A. G. NATHORST, "Zur fossilen Flora Japan's. Palæont. Abhandlungen herrausgegeben von DAMES und KAYSER." Vol. 4. 1888.

merely from the fact that a fourteen months periodic displacement of parts of seconds wanderings with an amplitude of 15 M. have been proved, but the data of palæontology are not in harmony with such a supposition. NEUMAYR was led to that supposition chiefly by the existence of a luxuriant Tertiary flora in Grinnell Land, and by the fact that the Tertiary flora of Alaska, at 60° N.L., scarcely bears the character of a southern flora any more than that of Spitzbergen at 78° N.L. does, and also that outside the polar circle the distribution of organisms seems to point to such a displacement. In fact, the Tertiary flora of the interior of North America bears the character of a subtropical or even of a moderately warm climate, and the Later Tertiary plants of Mogi in the Japanese island Kiu Shiu, described by NATHORST, point to a somewhat cooler

climate than the present. Neither could
DARWIN or PHILIPPI conclude from the
Tertiary marine animals found off the coast
of Chili, at 35° S.L. a temperature of the
sea water higher than the present comparatively cold one.*

It will no doubt be found striking that
all these regions have now also a comparatively cool climate. Since in the Tertiary
period the present continents (Europe ex-
cepted) had, at least in those parts of
North America, South America and Eastern Asia, almost the same outlines as they
have now, it will be more natural to seek
the cause of these Tertiary anomalies in

---

* CH. DARWIN, Geological observations on the volcanic
Islands and Parts of South America, visited during the
voyage of H. M. S. "Beagle", 3rd Edition, London 1891,
p. 413. The "Observations on South America" appeared
for the first time in 1846.

R. A. PHILIPPI, Die tertiären und quartären Versteiner-
ungen Chile's. Leipzig, 1887.

the general distribution of heat in influences which account for those anomalies nowadays, than in hypothetical considerable displacements of the earth's axis. The climate of an extensive continent like that of Tertiary North America, not directly influenced by warm ocean currents, must have been less fit for a subtropical vegetation than the insular climate of Europe, influenced as this was by warm currents, and the difference between the climates of these two parts of the world must have been still more considerable than it is now. It may have been that during the Pliocene period, when the general distribution of heat differed but little from that of the present time, the climate of Japan was somewhat cooler than it is now, if by the connection of the islands with the continent the climate was less insular and in that way the influence of the cold Oja-Shio

4

current on the east coast was stronger, while the influence of that branch of the warm Kuro-Shio current, which nowadays washes the west coast, would not exist.*

The west coast of South America owes its present relatively cool climate to the influence of the cold Peru current. As will be more fully explained in the second Part, the chief conditions of a greater intensity of all the ocean currents, and therefore of this current likewise, existed in the Tertiary period. Its greater power could thus compensate its relatively warmer origin, so that the sea water by which the Tertiary coast of Chili was washed was not warmer than it is at present ($14\,1/2^{0}$ C.).†

*On these influences acting on the present climate of Japan see: A. WOEIKOF, Bemerkungen über die Temperatur der ostasiatischen Inselreihe Sachalin, Jesso und Nippon. Zeitschr. für österr. Meteorologie. Vol. 20, 1885, pp. 1—3, and: Die Klimate der Erde. Jena, 1887. Vol. II, p. 365 sqq.

†According to E. SUESS (Das Antlitz der Erde. Vol. 2.

The cooler Tertiary climate of Alaska may also be explained by the same cause which renders it relatively cool at present: namely, predominating land winds. †

Moreover, a displacement of the pole would not suffice to render the Tertiary arctic phenomena any more intelligible to us. NEUMAYR supposes a displacement of the pole from 10° to 20° along the Ferro meridian in the direction of Eastern Asia. A displacement of 10° in that direction

Wien. Prag. Leipzig, 1888, p. 661) it is, moreover, still questionable whether the expression "Tertiary" applied to the respective marine deposits in Chili is really equivalent to the European term "Tertiary". H. ENGELHARDT (Abhand. der Naturf. Gesellsch. Isis. Dresden 1890, p. 3, and: Abhandl. der Senckenberg. naturf. Gesellsch. Frankfurt a. M. 1891) described Chilenian Tertiary plants—even from Punta Arenas at the Magellan Strait—which point to a damp tropical climate.

*W. H. DALL, Alaska Meteorology. Coast Pilot of Alaska. App. I. Washington, 1879.—Compare further: HANN's Isobarenkarte, No. 6 of the Abtheilung Meterologie in: BERGHAUS' Physikal. Atlas Gotha, 1887.

would certainly place Grinnell Land, Green land and Spitzbergen in more favourable conditions, but the Tertiary forests of New Siberia would then come even in closer proximity to the pole than Grinnell Land is at present. A larger displacement would lead to still more impossible results, and a much smaller one would produce no noticeable effect. The opinion expressed by NATHORST, that the Tertiary north pole has been situated on the meridian $120°$ eastern longitude of Greenwich, $20°$ more towards Asia, is in opposition to the discoveries lately made in New Siberia,* and also to those of the Tertiary florula of Punta Arenas,† which then would have existed at about $73\frac{1}{2}°$ S.L. NEUMAYR

---

*Compare: E. VON TOLL in: J. SCHMALHAUSEN, Tertiäre Pflanzen der Insel Neu-Sibirien, l.c., pp. 7—9.

† H. ENGELHARDT, Abhandl. der Senckenberg. Naturf. Gesellschaft, Frankfurt a. M. 1891.

himself must acknowledge that for any imaginary position of the pole those places where luxuriant Tertiary forests existed remain in much closer proximity to the pole than the present most northerly limit, where stunted trees are to be found, and that in any case the fossil flora of Europe deviates much more towards a warmer climate, than that of Japan towards a cooler climate than the present.

Should, therefore, astronomers and physicists raise no objections to a considerable displacement of the poles, the facts observed with regard to the arctic vegetation of the Tertiary period still offer sufficient arguments against such a supposition. The displacements of the earth's axis and of the poles, if any have taken place at all in early periods, can have been but trifling, for the climatic distribution of organisms during those periods which we have been able to examine in this respect, corresponds with zones which concentrically surround the present poles. *

We may thus conclude that, from the age of the first Palæozoic organisms up to the end of the Tertiary period, in high latitudes, either in the northern or southern hemisphere (temporary local anomalies not taken into account), warm climates prevailed, and that the deviation from the present conditions increased towards the poles. As far as we are able to ascertain,

---

* NEUMAYR himself considers it "a fact of great importance" that, according to his investigations, the limits of the homoeozic belts in the Jurassic period ran almost parallel to the present equator of the earth, which proves that "the equator and the poles cannot have undergone any considerable displacement since the Jurassic period" .... "the results, here obtained oppose conclusively the assumption that since the Jurassic period displacements in that direction large enough to account for considerable displacements of the zoogeographic zones have taken place." (Ueber klimatische Zonen während der Jura- und Kreidezeit. l. c. p. 307).

the regions near the equator have undergone but slight changes, and have (temporary local anomalies not taken into
account) never been colder but perhaps
slightly warmer than at present.

Gradual Tertiary Cooling of the Climates, and Pleistocene Glacial Epochs.

During the relatively short Tertiary
period a general cooling gradually took
place. Evidently under more favourable conditions, resulting from the distribution of
land and sea, this cooling commenced in
Europe somewhat later, but then proceeded
more rapidly than in North America. As,
however, in the latter continent the above-
mentioned local conditions underwent but
relatively little change from the commencement
of the Tertiary period, we must conclude
that the general cause from which this
cooling originated would act there least

disturbed, and that, precisely as has been observed in North America, this general cooling really commenced at the end of the Cretaceous period and lasted up to the end of the Tertiary period. The gradual decrease of heat going on during this whole period and the continual approach to the present climatic conditions find their unmistakeable expression in the change of the organic world, notably in its distribution. These will be generally acknowledged facts.

By the assumption that not earlier than the Tertiary period a considerable cooling of the previously almost homogeneous climate of the earth set in over the greater half of the earth's surface, an otherwise totally obscure chief feature in the palæontological history of the vertebrate becomes clear. For it is only at the end of the Mesozoic period that we see the reptiles,

the dominators of that long age, possessing
scarcely any heat of their own, and being
but poorly protected against changes of
temperature by their scales and shells,
recede rapidly before the classes of mammals and birds. Though these had already
appeared very early in single, small and
low forms, they first attained, and that
rapidly, a high flourishing state in the
Tertiary period. For these classes possess
in their warm blood and their bodily covering as well as in the development of
their eggs—which, unlike that of those
coldblooded vertebrates, requiring for this
development, as for their whole vital activity, external heat so much, is not directly dependent upon the temperature of
the climate—evident adaptations to a colder
and more changeable climate, which at
that time must have led them easily to
domination in the animal world.—The

intenser activity of the vital interchange
of matter in the homœothermal animals,
necessary to maintain the high and constant temperature of their bodies, was
the immediate origin of the circulation
of the blood and of the respiration, as
well as of the organs which effect a faster
and more thorough utilisation of food.*
In connection with this we observe that
the teeth of the mammals, which during
the whole of the Mesozoic period obtained
no higher degree of perfection than the
multitubercular type, already existing in
the Triassic period, from the beginning

*Compare: R. BAUME, Versuch einer Entwickelungsge-
schichte, des Gebisses. Leipzig, 1882, p. 284.

Mammals have on the average nearly a tenfold intenser
consumption of matter than reptiles and amphibians at
20° air temperature. At 1° air temperature this is in the
cold-blooded vertebrates nearly zero, and only at 36° is it
equal to that of the homœothermal animals. (Compare: L.
HERMANN, Lehrbuch der Physiologie. 9te Auflage. Berlin,
1889, pp. 215 and 212).

of the Tertiary period, develop rapidly, starting from the primitive Mesozoic trituber-cular type.—Again, the constantly intenser activity of the vital interchange of matter rendered possible a uniform production of a greater amount of external work, and facilitated the building up of the body whenever and wherever it was necessary. Thus the original requirement for the production of more heat caused the mammals and birds to attain higher functions and acquire more perfect forms than could be obtained by the poikilothermal reptiles; and even the high development of the brain in the mammal class may be considered as the result of that diminution of external heat which took place during the Cainozoic period.*

* LARTET and MARSH, it is well known, have really proved a gradual progress of development in the brain of mammals during the Tertiary period.

The Pliocene climate was but slightly warmer than the present, and at the commencement of the Pleistocene period the climatic condition of the earth was the same as that which prevails at present. Still later, a general cooling took place over the whole earth, the temperature falling below the present, and resulting in a general lowering of the snow line, which again caused enormous glaciers to descend from Scandinavia, Finland, and the Russian Baltic provinces, over a great part of Europe, as far as England, the Netherlands, Central Germany, and far in the South of Russia. A similar continental ice sheet covered North America as far as the 39th degree N.L. (the latitude of Lisbon in Europe), Great Britain, the Alps, the other high mountain ranges of Central Europe, the Pyrenees and the Apennines, the Caucasus, the Atlas in North Africa and even, though in less degree, the high mountains in the dry regions of Northern Asia: also the Himalayas, the Karakorum and Tien Shan mountains. In the southern hemisphere, Patagonia, the Andes of Chili, perhaps also the higher mountains of South Africa, the Australian Alps and especially New Zealand, also the Kerguelen, the South Georgia and Falkland Islands had their own gigantic system of glaciers, partly extensions of the present ones, and partly existing on the higher mountains which are not ice-covered at present. All these regions are still covered with mineral matter transported by glaciers. Nearer to the equator traces have also been found of Pleistocene glaciers in the Sierra Nevada de Santa Marta (at 11^0 N.L.) and the Andes of Cocui ($4\frac{1}{2}^0$ to $7\frac{1}{2}^0$ N.L.) in Columbia, and in the Sierra Nevada de

Merida (8° N.L.) in Venezuela.* These occurrences of Pleistocene Glacial phenomena near the equator are one of the best evidences that the glaciation of the northern and southern hemispheres took place simultaneously, so that its general character can no more be doubted.†

This general depression of the snow line was accompanied by a generally estab-

---

*W. Sievers, Ueber Schotterterrassen, Seen und Eiszeit im nördlichen Südamerika. Geogr. Abhandl. herausgegeben von A. Penck. Wien 1887. Vol. 2.—A. Hettner, Reisen in den Kolumbianischen Anden. Leipzig 1888, p. 319, and: Die Kordillere von Bogotá. Petermann's Mitteilungen. Ergänzungsheft No. 124. 1892, p. 75.

†About this general and simultaneous prevalence of the Glacial phenomenon see: A. Penck, La période glaciaire dans les Pyrénées. Bull. Soc. d'hist. nat. Toulouse. 1885. Vol. 19, p. 107. Compare also his compendious description of the entire Glacial phenomenon according to the present state of our knowledge: Die grosse Eiszeit in: Himmel und Erde. 4. Jahrgang. 1891, pp. 1 and 74.

Beyond the above mentioned pleistocene glaciations in the tropics, the following ones have been found, which I mention

lished increase of lakes without effluent. *
As the dampness of the climate is a
still more important factor in the formation
of glaciers than a low temperature, it has
been suggested that during the Glacial
epoch the temperature might not necessarily have been lower, but might even
have been higher than at present. † Yet

here on account of the importance of this subject's bearing;
in the Andes of Ecuador (El Altar) near the Equator, in the
Ilimani and the Andes of Bolivia between 15° and 18 ¹/₂° S.L.
in the Anahuac of Mexico as far as 17° N.L. and in the
Namuli mountains of East-Africa at 16° S.L. The glaciation
of the Sierra do Mar in Brazil asserted by AGASSIZ and
HARTT is according to J. C. BRANNER (A. R. WALLACE,
Darwinism, London, 1890, p. 370) decidedly problematical.—
Compare also different writings of PENCK, especially his latest
essay in: Himmel und Erde, and: Map. No. 5 of the geo-
logical part of the new edition of BERGHAUS' Physikalischer
Atlas.

*E. BRÜCKNER, Klimaschwankungen seit 1700, nebst Be-
merkungen über die Klimaschwankungen der Diluvialzeit.
Wien 1890. Cap. x.

†J. D. WHITNEY, The climatic changes of later geological

not only have serious objections been raised by meteorologists against this suggestion,* but also the palæontological facts of the Glacial epoch are in no wise in harmony therewith.   Not only was there a depression of the snow line, but also of the forest line, which certainly is only dependent upon the temperature. The entire flora and fauna of the glaciated regions belong to a cool climate,† and cannot in the least be compared to the luxuriant life of Patagonia and New Zealand where, favoured

times. Mem. of the Museum of Compar. Zoology at Harvard College. Vol. 7. No. 2. Cambridge, U. S.'A. 1882.

* A. WOEIKOF, Die Klimate der Erde. Part. I, p. 102 sqq.

† A recent treatise on the distribution of Arctic plants during the Glacial epoch in Central Europe, by A. G. NAT-HORST, appeared in: Nature, Vol. 45, 1892, pp. 273—275: Fresh evidence concerning the distribution of Arctic plants during the Glacial Epoch.—The Arctic mammalia which are known at present from the Diluvian beds of Central Europe are mentioned by A. NEHRING, Ueber Tundren und Steppen der Jetzt- und Vorzeit. Berlin 1890, p. 167.

by a relatively warm, but especially uniform and damp climate, a subtropical vegetation can exist even in between the glaciers.

In calculating approximately the amount of melting (ablation) at the known snow line of the Glacial epoch with the distribution of heat, supposed to have been the same as now, this is found to be three or four times as considerable as it is at the present snow line. During the Glacial epoch the precipitation ought, therefore, to have been three or four times as much as it is at present; which is certainly impossible. * The former supposition that over Europe Arctic cold then prevailed has long been given up, still everything indicates that the temperature must have been several degrees less than the present. † Judging from the

* A. PENCK, Die grosse Eiszeit, l.c., p. 13.

† This conclusion has been arrived at principally through climatic investigations of A. WOEIKOF, Ueber die klimatischen

character of the flora and fauna, and from the extension of the ice, we may infer that the difference from the present condition must, at temperate latitudes, have been less than 5⁰ C. PENCK succeeded in computing approximately the height of the pleistocene snow line on a number of mountains, and found it to have been on the average about 1000 metres (from 500 metres to 1300 metres) lower than the present, which corresponds with a general reduction of the earth's temperature of 5⁰ C.* He also calculated that during the Glacial epoch such a considerable quantity of water was withdrawn from the ocean, that its mean level must have been about 100 metres lower. "A general reduction

Verhältnisse der Eiszeiten sonst und jetzt. Verhandl. d. Gesellsch: für Erdkunde. Berlin 1880. Vol. 7, p. 151 sqq.

*A. PENCK, Die Eiszeit in den Pyrenäen. Mitteilg. d. Ver. f. Erdkunde. Leipzig 1884, p. 209.

of temperature of 4°, 5 C.—and for a climate damper than the present, of even less—suffices to account for the whole Glacial phenomenon."*

The fauna of the North Sea had an arctic character. TORELL stated that the marine mollusca and the marine mammals, which are found in the *Yoldia* clays of the Cattegat and the North Sea, indicate temperatures which prevail at present in the Karian Sea.† In all probability, therefore, the temperature of the sea water in that part of Europe was, during the coldest period, 5° to 10° C. lower than the present; which is quite natural when we take into account that huge masses of northern inland ice descended into those seas. The Mediterranean Sea at that

---

* A. PENCK, Die grosse Eiszeit, p. 77—79.

† O. TORELL in: Zeitschrift der Deutschen Geologischen Gesellschaft. 1888. Vol. 40, p. 250 sqq.

period, also contained many organisms which now exist in the North Sea. To judge from the organisms, the difference of temperature, at least for the sea climate, must have decreased towards the equator.

This cold, which prevailed over the whole earth, did not consecutively last up to the beginning of our time. In Europe and North America, where the Glacial phenomena have been longest known and most fully investigated, an intermediate bed, which shows no trace of ice agency and contains remains of plants and animals indicating a climate no cooler than the present, has been found in many places between Glacial formations. These Interglacial formations prove that during the Glacial epoch a recurrence of the temperature which prevailed at the beginning of the Diluvian age intervened, during which,

the great ice masses dissolved and a
flora and a fauna existed similar to
that of the Preglacial and Postglacial (the
present) epoch but indicative of a drier
climate during the acme of the epoch
(Steppe periods). * A second, older, shorter and less conspicuous Interglacial epoch
has been found in Europe, which, nevertheless, must have lasted much longer than
the Postglacial epoch, in which we live. †

* Compare: F. VON RICHTHOFEN, China. Vol. 2, p. 348,
and: Ueber die Bildung des Löss. Verhandl. der KK. Geol.
Reichsanstalt in Wien. 1878, pp. 289—296, and different
treatise on Steppenfaunæ by A. NEHRING, chiefly his recent
comprehensive treatise: Ueber Tundren und Steppen der Jetzt-
und Vorzeit. Berlin 1890.

† A. PENCK, Zur Vergletscherung der Deutschen Alpen.
Leopoldina. Vol. 21. Halle 1885, pp. 105, 129 and 145.—
E. BRÜCKNER, Die Vergletscherung des Salzachgebietes, etc.
Geogr. Abhand. edited by A. PENCK. Wien 1886; I, p. 133,
and: Die Eiszeit in den Alpen. Mitteil. Geogr. Gesellsch.
Hamburg 1887—1888, pp. 10—23, and further his: Kli-
maschwankungen. Cap. x.

The shorter duration of the last mentioned epoch is as apparent from the relative thickness of the Interglacial and Postglacial Weathering-Loam as from the Interlacustre and Postlacustre deposits in the North American lake basins of the Diluvial period investigated by GILBERT and RUSSELL. The penultimate Ice Age was of longer duration than the last one, and the oldest known was the shortest.

Thus the Glacial phenomenon consists of a general cooling of only a few degrees, a cold which differed still less from the present temperature than the higher temperature of the Pretertiary period; it occurred certainly twice and probably three times. During the long intervals which separate these cold epochs the same climatic condition recurred which obtained earlier and later and still prevails now. These cold epochs must, therefore, be considered as interruptions of the present condition, which set in as early as the beginning of the Pleistocene period, immediately after the Tertiary period.

PART II

Changes in the Solar Heat as the Agency by which the Geological Climatic Changes were brought about.

Considerable changes of the solar radiation sufficient to account for the climatic changes.—The postulated changes of solar radiation have actually taken place.—The solar heat arises most probably from contraction. Seeming inconsistency with geological and biological data.—Further circumstances in favour of the view that the geological changes of climate are to be explained by the history of the sun's evolution.

Considerable Changes of the Solar Radiation sufficient to account for the Climatic Changes.

In the preceding Part the general features of the great changes in the heat on the

surface of the earth, according to the palæontological and geological data, were discussed.
What can have been the cause of these
changes? As it has now been sufficiently
proved that tellural agencies (internal heat of
the earth, a possible different mixing of the
atmospheric gases, changes in the division
of land and sea),* if they prevailed at all,

---

*The belief is still too common that ages have been, during
which heat rising from the interior of the earth could have
contributed in a high degree to the development of life on the
earth; though as early as 1862 Sir WILLIAM THOMSON (now
Lord KELVIN) proved "that the general climate cannot be
"sensibly affected by conducted heat at any time more than
"10,000 years after the commencement of superficial solidification." (On the secular cooling of the earth. Trans. Roy.
Soc. Edinburgh. Vol. 23. § 17, also published in his: Mathe-
matical and Physical Papers. Vol. 3. London 1890, p. 305).
That solidification must have commenced more than 20,000,000
years ago (Mathemat. and Phys. Papers. Vol. 3, p. 300).

I need not dwell here upon the astronomical hypotheses
of CROLL, BALL, and others for explaining the geological
climatic changes. These, as it seems to me, have been con-
demned sufficiently, though they still find many adherents,

could never have sensibly modified the general heating of the earth during the time it has been inhabited; and it being further proved that the more uniform distribution of heat up to the Tertiary period was such that, though places of high latitudes were favoured by it, this

---

chiefly in England. (Compare: A. WOEIKOF, Examination of Dr. CROLL's Hypotheses of geological climates. American Journal of Science. 1886. (3) Vol. 31, p. 161, and: Philosoph. Magazine. 1886, p. 233 sqq.) Yet it must be mentioned that, as NEUMAYR (Erdgeschichte Vol. 2, p. 649), draws attention to, the climatic conditions of the Diluvian period, on which these explanations are based, form but an exceedingly small part of the climatic changes which occurred during untold millions of years. "Notwithstanding all study it is, "therefore, not possible to form a just opinion from this "exceedingly limited space of time. We must first try to "know the climatic conditions of the former periods".... And, certainly, neither the development of the organic world, nor its distribution in the former ages, agree with the assumption of winter cold then already existing and recurrent glacial and interglacial periods, for the assumption of which geological data likewise do not plead.

was not at the expense of the equatorial regions, it is obvious that the total amount of heat received by the earth must have been greater than at present. In fact it must have been considerably greater; for in January the earth receives $6\frac{1}{2}$% more heat from the sun than in July and this difference not only nowhere produces any noticeable effect, but its general temperature in July is even higher than in January.* The influence of the larger share of solar heat, which the earth receives in winter time, does not appear to be felt through the thick atmospheric mantle which is so changeable in its motion, the amount of vapour it holds and its clouds. The unequal distribution of land and sea in the northern and southern hemispheres—an

---

*R. SPITALER, Die Wärmevertheilung auf der Erdoberfläche. Denkschr. Math. Nat. Kl. K. Akad. der Wissensch. Wien 1886. Vol. 51, p. 1.

unimportant factor in the general diffusion of heat over the globe, as has been proved by geological experience*—influences the general temperature more than this increase of one fifteenth of the solar heat.

In the, geologically speaking, recent period, during which the present diffusion of heat prevailed, it has twice happened that the earth received a considerably smaller quantity of heat than at present, as otherwise there would not have been a marked influence upon the climates.

It now seems to be completely proved that no other source of heat than the sun can ever have exercised an appreciable influence either on the meteorological condition of the earth, or on the climates. To any very considerable changes of the solar heat we must, therefore, look for the cause

* Chiefly by the valuable investigation made by NEUMAYR, Die geographische Verbreitung der Juraformation. l.c.

of the geological changes of climate. But the question then arises whether, assuming a considerable diminution or increase of solar radiation, the diffusion of heat over the surface of the earth would be like what we know that it must have been from the geographic distribution of organisms!

This must really have been the case, as is proved in the first instance by the present diffusion of heat. Comparing the various temperatures for different latitudes, as computed by ZENKER * from the solar radiation, with those found by SPITALER † from observation (Table xxii of ZENKER), a considerable displacement of heat from the equator towards the north pole becomes apparent.

---

*W. ZENKER, Die Vertheilung der Wärme auf der Erdoberfläche. Berlin, 1888.

†R. SPITALER, Die Wärmevertheilung auf der Erdoberfläche, l.c.

Between 50⁰ and 70⁰ northern latitude a great excess of heat is found, which corresponds with a deficiency in the equatorial regions. In the transition zone, from 10⁰ to 50⁰ northern latitude, that excess is only very slight. Its origin must be found, as may be seen from the maps of thermal isonomals of SPITALER,* in the warm ocean currents and in the predominating winds. Thus the yearly temperature in the Atlantic Ocean north-west of Norway is about 12⁰ C., and in January even 24⁰ C. higher than it ought to be theoretically, according to the latitude. This same influence is still very marked in the temperature of Spitzbergen and even of Greenland and Nowaya Zemlja. Comparing the land temperatures at the equator, as computed by ZENKER and also

* Ibid.—Compare: Petermann's Mitteilungen 1887. Taf. 20 and 1889. Taf. 17 and 18.

by FORBES and SPITALER,* with that which has been actually observed, a great difference is apparent, the latter being 10 and more degrees Celsius lower; the observed sea climate of the equator, on the contrary, being nearly the same or even a little warmer than the computed one. Hence we may conclude that not a small amount of the excess of heat in high latitudes is due to the heat received by the land in equatorial regions. The southern hemisphere is cooler than it ought to be according to calculation, which must ultimately be attributed to the fact that it consists mainly of sea. † As it is known, however, that the northern hemisphere had even in

---

*W. ZENKER, l.c. Tab. xxiii. J. D. FORBES, Inquiries about terrestrial temperature. Trans. Roy. Soc. Edinburgh. Vol. 22.—R. SPITALER, Die Wärmevertheilung auf der Erdoberfläche. l.c.

† Compare: Chapters VI and XVI of Vol. I of: WOEIKOF, Klimate der Erde.

early ages a relatively large proportion of
land, and as we judge of the temperature
of early climates mainly by land organisms,
we may safely deduce from the results
obtained by ZENKER for this hemisphere
in its present condition its climatic conditions in former periods.

Changes in the solar radiation must,
according to ZENKER, have three times as
strong an effect on the computed (solar)
land temperature of the equatorial regions
as on temperate and polar latitudes, the
solar land temperatures increasing in proportion to the augmentation of the yearly
solar radiation. The excess of heat at
high latitudes must, therefore, increase with
an increase of solar radiation and must
have embraced a broader zone than at
present, owing, as now, to the heat received by the sea at the equator—which
heat, as will be more fully explained here-

after, would then be yet more easily transported because of the more rapid circulation of the atmosphere.

Owing to the extensive area of the tropical belt, 40% or two fifths of the total area of the earth's surface—which embraces an area almost as large as the two temperate belts, and four times as large as the two polar caps—and the large amount of heat received, even small changes in its temperature must exercise a great influence on the climate of high latitudes.

This influence is mainly exercised by the agency of ocean currents, caused by the prevailing winds (which themselves probably play an important direct part in the distribution of heat). But even though the sea area within the tropical belt should have been much smaller than the present (three fourths of the total area), the influence of the equatorial heat on the higher

6

latitudes would not necessarily have been less, as these latitudes profit so greatly by the heat received by the land at the equator. The direct result of an increase of land area at the equator would, therefore, be an increase of the excess of heat at high latitudes, though this increase would have been compensated by the decreased quantity of heat transporting water.

Broad communications between the seas at high latitudes and the tropical seas, and connections, such that the warm water could flow through in the direction indicated by the predominating winds, must, however, have been necessary conditions for the influence described above. We know with certainty that in the Carboniferous, the Jurassic, the Cretaceous and the Tertiary periods, of which the arctic floras are known, such broad connections, even much broader than the present, existed between the arctic seas and those near the equator or in the tropics.

An increase of slow radiation must in the first place have *directly* raised the temperatures of temperate and high latitudes, chiefly in favour of the land temperature—but a far greater influence must have been exercised *indirectly* by the heat transported from the tropics by warm ocean currents by which the sea climate would chiefly be benefited.

Diminution of the sun's heat would, on the contrary, have had the effect, that in high latitudes the indirect cooling of the sea was greater than the direct one which had mainly affected the land.

At an age in which the surface of the southern hemisphere consisted, as at present, mainly of water, the changes of the heat

received there must have been partially
felt in the northern hemisphere: in the
same manner as at present a great deal
of the warm water of the southern hemisphere flows into the northern hemisphere.
The distribution of land and sea was, at
least during the Pleistocene period,
very probably the same as it is now,
and thus the smaller supply of warm
water to the northern hemisphere must
have been a cause of refrigeration of that
hemisphere.

A much warmer sea washing the shores
of a still warmer continent, and a more
considerable exchange of heat in the
direction of the meridians, caused by the
stronger general system of winds—this
was, therefore, the climatic condition in
higher latitudes under the reign of a hotter
sun—a cool sea, more especially in
the northern hemisphere, together with a

slightly cooler continent, must have been the effect of a less hot sun. A similar condition in which the seas are cooler than the land obtains nowadays in Patagonia and New Zealand, where glaciation is stronger now, than anywhere else at a temperate latitude. Such a condition is the most favourable for glaciation and we may safely conclude, with J. D. FORBES and WOEIKOF,* that by a considerable cooling of the seas between 40° to 60° N.L. many mountains in Europe and in the western part of North America would be covered by glaciers and that the existing glaciers would increase.

It is quite possible that in the tropics, where a greater dampness must have prevailed during the Pleistocene period, even slight diminutions of the land temperature, which probably have taken

* A WOEIKOF, Die Klimate der Erde. I. p. 104.

place, caused a considerable increase of the present glaciers and originated new ones.

There is now every reason for assuming that actually, under the influence of a hotter or less hot sun, the ocean currents, which can exercise in the diffusion of heat over the surface of the earth so great an influence,* and the prevailing winds, which are the cause of the sea currents, and apparently also exercise a strong direct influence, † ¨have had, as a rule the same direction as now, and that

---

*The works of CROLL are still fully deserving of attention in this respect, chiefly: Climate and Time in their Geological Relations. London 1875, and his earlier essay: On Ocean Currents in Relation to the Distribution of Heat over the Globe. Philosoph. Magazine, 1870.

† H. VON HELMHOLTZ has in his papers: Ueber atmosphärische Bewegungen. Sitzungsber. der K. Pr. Akad. d. Wiss. Berlin 1888, and: Meteor Zeitschr. 1888, Vol. 5, pp. 329—340, pointed out that by the displacement of the air masses, from the equator in a northerly and southerly direction,

also the force of these winds varies considerably according as the sun's heat in more or less powerful. More and more, nowadays meteorologists recognise the fact that a grand system of circulation prevails in the atmosphere of the earth, between the equator and the poles, of which the predominating winds form part. Chiefly by the admirable theoretical investigations of FERREL and WERNER SIEMENS (who, starting from entirely different bases, arrived at the same main conclusions), * as well as by the observation of atmospheric pressure and temperatures

an enormous amount of kinetic energy is used, which is transformed into heat.

*As early as 1860 W. FERREL proclaimed the existence of a general circulation. More within the reach of everyone than his earlier publications is his: Popular Treatise on the Winds. London 1890.

WERNER SIEMENS, Ueber die Erhaltung der Kraft im Luftmeere der Erde. Sitzungsber. d. K. Pr. Akad. d. Wiss. Berlin 1886. Vol. 13, p. 261.

at mountain stations, and their interpretation by HANN, it is now a recognised fact
that a general circulation prevails in the
atmosphere, dependent either upon the
rising and then poleward directed force of
the heated air at the equator (SIEMENS) or
from the constant difference of temperature
between the poles and the equator,
namely the thermal gradient (FERREL). This
general wind system consists on the surface of the earth of two large whirls, one
in either hemisphere, moving according to
the law of BUYS BALLOT, and an equatorial
wind belt moving in an opposite direction.
It produces at the surface of the earth at
temperate and higher latitudes, from the
35th degree, in the northern hemisphere
south-westerly winds and in the southern
hemisphere north-westerly winds (westerly
winds), the directions of the winds near
the equator being north-east in the northern

hemisphere and south-east in the southern hemisphere (trade winds). At the 35th degree of latitude (calm belts of Ross) there is an increase in the atmospheric pressure, which is caused, either by the " centrifugal force" (FERREL) or by the forcing up of the poleward upper current in consequence of the narrowing of the bed of the current (SIEMENS). According to SIEMENS—and his theoretical investigations have been inductively affirmed by HANN*—cyclones also owe their origin to this general circulation; the current, which, by the narrowing of

---

*See the most recent paper of this eminent meteorologist: J. HANN, Das Luftdruck-Maximum von November 1889 in Mittel-Europa, nebst Bemerkungen über die Barometer-Maxima im Allgemeinen. Denkschr. Math. Nat. Kl. d. K. Akad. d. Wiss. Wien 1890. Vol. 57, pp. 401—424, and: Studien über die Bedingungen von Luftdruck und Temperatur auf der Spitze des Sonnblick, nebst Bemerkungen über ihre Wichtigkeit für die Theorie der Cyclonen und Anticyclonen. Ibid. 1891. Vol. 58.

its bed, is continually retarded on its way
towards the pole, and which by the earth's
rotation is continually deviated towards
the last, draws the lower atmospheric
layers along by internal friction, thus pro-
ducing local minima.

This general circulation of the atmosphere
will intensify increases with the difference between the heating of the equator
and the poles—as is the case in every
augmentation of solar radiation—and will
diminish with a decrease of that difference—
as in every diminution of solar radiation.
Moreover, in the first case the meridian
component of the wind's direction (that is,
the one which is dependent upon the
amount of heating) will be predominant, in
the latter case the parallel one (that which
corresponds with the earth's rotation).
Thus, at a time when the earth receives
more heat, the winds on the polar sides

of the 35th degree N. and S. latitude
not only increase in force, but have
also a more northerly direction; while
at a time of less heating of the earth
their direction must be still more easterly
than under the present conditions.   Further,
the ocean currents, which owe their origin
to this general wind-system, are, with regard
to their power, velocity and direction,
subject to the same changes and will
transport, in the first case much more, in
the latter case much less heat from the
equator towards the poles.   As the main
directions of the oceans in the Diluvial
period were apparently the same as at
present, then the continents must have had
in the latter case—with more easterly
directed winds—also a damper climate
(corresponding with Glacial periods), and
in the periods of moderately increased solar
radiation a drier climate with winds of a

more northerly direction (Interglacial Steppe periods). But at the time of a greater increase of radiation intensity, the result of the prevailing winds setting in a more meridianal direction would certainly not equalize their greater increase of force and quantity of vapour—at least where predominating influences otherwise acting did not exist, as seems to have been the case in Europe during the Eocene period.*

The influence of the reduced force of ocean currents (and of predominating winds) during the Ice age is apparent from the fact that, during that period, Europe lost the climatic advantage which it enjoys at present over the east coast of North America. The British

---

* According to DE SAPORTA the flora of the European Eocene was that of a warm dry country with little rainfall.

seas in particular suffered greatly in this respect.*

Further we observe that the system of westerly winds increases in force and sets in at lower latitudes during the winter, when the difference between the temperature at the equator and at the pole of the relative hemisphere is greater than in summer. This temporary increase of the force of the predominating winds can have no perceptible effect on the ocean currents, which are the result of the entire work performed for centuries by those winds. Their moment of motion is much smaller than that of the ocean currents,† and their immediate effect, therefore, almost

---

*Sir J. W. DAWSON, The Story of the Earth and Man. 10th Edition. London 1890, p. 273.

†G. VON BOGUSLAWSKI and O. KRÜMMEL, Handbuch der Oceanographie. Stuttgart 1887. Vol. 2, p. 351. The proportion of the moment of motion of air currents to that of ocean currents is according to KRÜMMEL as 1 : 123.

imperceptibly small; the stationary state of motion of the ocean, corresponding to the average velocity of the winds, has obtained for thousands of years, and the winds have only now to make good the loss caused by friction, to which their moment of motion certainly suffices.* For this reason the stronger circulation of the air during the winter of either hemisphere cannot serve to raise its temperature perceptibly, for by the smaller specific heat of the air, in comparison to that of land and water, it will not reach temperate latitudes before being entirely cooled in its long journey from the equator, and this winterly strengthening of the movement of the air is not yet sufficiently strong to exercise a considerable influence by its kinetic energy becoming heat.

* Ibid., p. 350. Conclusion of HANN.

This direct influence of the winds towards an equalisation of the differences between the temperature at the equator and at the poles must certainly have been greater during the time of a stronger circulation,* but even at that time it must still have been inferior to the influence exercised by the ocean currents whose specific heat is so much greater. Assuredly this influence of the winds must—then, as now—have contributed more towards equalising the diffusion of heat over short distances, between neighbouring seas and lands.

That the direct influence of changes in the solar radiation—though according to ZENKER three times as great near the equator as in temperate and polar latitudes—

---

*According to the investigation of H. VON HELMHOLTZ (Ueber atmosphärische Bewegungen. ll. cc.) the direct influence of a considerable increase of the general circulation must have been very important.

does not, actually, in the first place affect the temperature at the equator, but the general circulation of the atmosphere, has been proved by the relation which exists between the eleven year sun-spot periods and meteorological phenomena. During the maxima of sun-spot periods the solar radiation is certainly largest. This is proved, in addition to spectroscopic observations, * in the first place by the fact that, according to H. F. BLANFORD, the atmospheric pressure then decreases at the earth's surface between the tropics (South Eastern Asia) and

* The surface of the photosphere covered with faculæ, which displays a noticeable increase of the continuous part of the spectrum, is much greater than that of sun-spots (generally increasing and decreasing parallelly with the faculæ), and these spots, according to their spectrum and to direct measurements, emit less heat than the main photosphere. Further, the spectrum of a sun-spot in the time of minimum seems to indicate a lower temperature than that of a sun-spot of maximum period (S. J. PERTY and A. L. CORTIE, Monthly Notices Roy. Ast. Soc., 49, 1889, p. 410).

increases at higher latitudes (Western Siberia and Russia in winter), apparently owing to an increase of the difference in heating between the equator and the poles, which must be accompanied in the upper regions of the atmosphere by an increase of the surplus of atmospheric pressure at the equator, which, again, causes an increase of the general wind system; * in the second place by the fact that, during the maxima of sun-spot periods, cyclones are most numerous in the tropics and, on the contrary, are scarce during the minima of sun-spot periods (C. MELDRUM, A. POEY); and in the third place by the fact that probably the rainfall is most abundant

---

*E. BRÜCKNER also pointed out, that for the 35 year periods of climatic changes—which he attributes likewise to oscillations in the intensity of solar radiation—the difference between the atmospheric pressure at the equator and at subpolar latitudes increases in warm periods, and decreases in cool periods.

during the maxima of sun-spot periods, which points to an increase of evaporation and an increased motion in the atmosphere. Still, the temperature of the air near the equator is generally lowest in maxima of sun-spot periods and highest in minima of sun-spot periods (W. KOPPEN, H. F. BLANFORD.)

All these phenomena make it apparent that the atmosphere operates as a very active regulator in the diffusion of heat over the surface of the earth, and that this activity is decreasing or increasing with the amount of heat emitted by the sun. As changes in the quantity of heat received can exercise directly so much more in-fluence in the tropics than near the poles, and consequently must be accompanied by a considerable increase or diminution of the general circulation of the atmosphere, they will only be of small influence on the actual temperature of the air at the equator, but, on the contrary, indirectly, of very great influence upon the temperature at high latitudes.

The postulated Changes of Solar Radition have actually taken place.

As considerable changes in the solar heat must consequently apparently suffice to explain the observed climatic changes, the question arises whether they have really taken place.

This is actually the case and by these changes we can satisfactorily explain the known main phenomena of the geological climates.

It has been repeatedly remarked, by the way, that, as the sun is constantly losing its energy, it must once have been hotter than at present and may thus have produced the higher temperature of the early geological ages. *

As a general remark this, certainly, is perfectly correct; but a gradual cooling down during the entire geological time has not been observed in the climates of past ages. During the long Jurassic and Cretaceous periods we see no marked change with regard to the general conditions of the heat over the earth, and it is now accepted that the temperature in temperate latitudes can have been but little higher

*F. PFAFF, Geologie als exacte Wissenschaft. Leipzig 1873, pp. 38 and 39.

Sir WILLIAM THOMSON, Transactions Geol. Soc. of Glasgow. Vol. 5, 1877, p. 238.

J. D. DANA. Manual of Geology. Third edition. New York and Chicago. 1879, p. 716.

A. GEIKIE, Text-Book of Geology, Second Edition, London, 1885, pp. 19 and 21.

Compare also: J. J. MURPHY, The Climates of Past Ages, in: Nature, Vol. 42. 1890, p. 270, and: E. BRÜCKNER, Klimaschwankungen, p. 315.

during the Palæozoic age than during the
latter part of the Cretaceous period. On
the contrary the general coolings must
have progressed very rapidly during the
relatively short Tertiary period.

As far as I know no attempt has as yet
been made to compare the sun's history
in full detail, as it must be according
to the now almost generally accepted and
everywhere so brilliantly confirmed theory
of KANT and LAPLACE, with the history of
the geological climates. Yet it appears to
me that, from the present state of science,
these last find a satisfactory explanation in
the history of the sun's energy.

The sun is a star like the thousands
which appear at night as luminous specks
in the celestial vault, even of relatively
small dimensions and differing from the
others merely by its closer proximity to
the earth. The substances of which it

consists are entirely the same as those of which many other stars are composed. The spectrum of the yellow stars, belonging to the second class of VOGEL, in every respect agrees with that of the sun, and it may, therefore, be confidently affirmed that the sun is a yellow star, and that the other yellow stars not only have the same composition, but also an almost equally dense and hot atmosphere as our sun.* From the colour and the spectrum of the stars we not only learn their composition but also their temperature,† for as soon as, at a certain temperature, a body begins to glow, the luminous and chemical rays appear. The higher the temperature rises, the richer the spectrum becomes in blue and violet rays; the lower the temperature

*A. SECCHI, The Stars.

†J. JANSSEN, L'âge des étoiles. Ciel et Terre. 2e Série. 3e année. 1887—1888, p. 465 sqq.

falls, the more the less refractible yellow and red rays will predominate. A star whose spectrum is very rich in violet rays will, therefore, at least in its external parts, possess a very high temperature. The number of the stars belonging to this class, of which Sirius, the most beautiful of all, is the most brilliant, but Regulus the most typical representative, is very great. More than half—58.5%—of all known stars belong to the first class, that of the white (bluish white) stars. They possess a dense, hot atmosphere of hydrogen, as seen by the four characteristic broad lines in their spectrum; we also observe in many stars of this class a number of almost imperceptibly thin lines, indicating the presence of metallic vapours, chiefly of iron.

The stars of the second class, to which our sun belongs, and with Capella as a

typical representative, possess no longer that,
bright white colour, but are yellowish
some even orange. From this we may
infer that they are less hot. Instead of
a mighty atmosphere of hydrogen, the
spectroscope shows us here, by very
numerous thin lines, the same metallic
vapours which we observe in our sun.
Hydrogen lines are still present, but they
are very thin and not at all so conspicuous
as in the white stars. This second class
is still very numerously represented, it
contains 33.5% of all visible stars.

Other stars of a red hue, of which
Beteigeuze is a type, are in a still later
stage of evolution; their spectrum indicates
unmistakably an already far advanced cooling.
The violet has entirely disappeared from
these broad dark bands or columns covering
the lines, similar to, but stronger and
more numerous than those of the second

class.   To this class belong about 8°/o of all the stars.*

* These figures, indicating the proportion in which the stars are divided in the three spectral classes, are computed from the Potsdam spectroscopic "Durchmusterung" (Public. d. Astrophys. Observat. zu Potsdam. Vol. 3. Comp.: J. SCHEINER, Die Spectral-analyse der Gestirne. Leipzig 1890, p. 325).

E. C. PICKERING (Preparation and Discussion of the DRAPER Catalogue. Annals of the Astronomical Observatory of Harvard College. Vol. 26. Part 1. 1891) has lately arrived at a different conclusion, by way of photography. He found for the figures indicating the proportion in which the three types of star spectra are divided 75, 23 and 1 (p. 151).— But the photographic plate is in this classification much more one-sided than the eye, which is most sensitive to the rays of the stars belonging to the middle class. Feeble stars of the third class are, therefore, photographically never to be distinguished from those of the second class (l. c., p. 178), just as the classification is difficult for all bright stars of less pure type, such as α Tauri and α Orionis, and all the less bright stars (Op. cit. Vol. 27, Preface). The Classification according to direct eye observations will therefore certainly express the real proportion more accurately.

Taking into account the one-sidedness of the photographic plate (being chiefly sensitive to the rays of the violet side of

Relatively few stars form the transition from the first to the second class; relatively more numerous are those which, according to their colour and spectrum, must be ranged between the second and the third class and therefore form the transition from the second to the third stage.

Except the classes of stars mentioned, which we must assume to have been developed out of one another (and the few stars which cannot be ranged in these three classes), we further know, as the

the spectrum) by the arrangement of the spectral types, it does not appear surprising that, as found by J. C. KAPTEYN, (See: Proceedings of the Royal Academy of Sciences at Amsterdam on April 29, 1892 and his Memoir on Stellar Magnitudes in Relation to the Milky Way, published in the Bulletin du Comité International permanent pour l'exécution photographique de la carte du ciel, 1892) the number of the stars of the first type—according to the PICKERING photographic classification—in relation to the number of that of the second type, increases in proportion to the distance from us.

first stage in the evolution of the luminous
celestial bodies, the nebulæ, consisting of
very thin and voluminous gaseous masses,
which, according to the now current
hypotheses, originated by the falling of dark
bodies upon one another. In these nebulæ
only hydrogen has as yet been proved to
exist with certainty. This class is so
limited in number that it contains less
thousands than the stars contain millions.

Besides all these visible bodies many
dark bodies (planets and extinguished
suns) must exist,—the number and the
total mass of which must even considerably exceed those of the visible
bodies.

This universe has not existed for ever,
but must have had a beginning. If not,
the forces of nature would have attained
an equilibrium and we should not be able
to observe nature, as has been pointed

out by FICK * and SECCHI. † But the universe which is known to us is not infinite. If this were the case, says SECCHI § with OLBERS and CHÉSAUX, we ought to see the celestial vault in its entirety as luminous as the sun, and as this is not so, we are forced to the conclusion that the number of stars cannot be infinite. SECCHI, however, acknowledges that doubtless a great many dark bodies exist in space which might intercept the light, but by the great distances of the luminous, as well as of the dark bodies, the latter would, like the dust in our atmosphere, be able to weaken the light, but not to entirely absorb it. If these dark

* A. FICK, Die Naturkräfte in ihrer Wechselwirkung. Würzburg 1869, p. 70.

† A. SECCHI in a lecture delivered March the 6th 1876 at Rome on the Greatness of Creation with regard to space and time.

§ A. SECCHI, The Stars.

bodies are very numerous, as they really must be, their intercepting and weakening of the light may be so considerable that the whole background of the heavenly vault may be but very slightly lighted, and form a dark screen against which the stars, which we are nearest, will sparkle like luminous specks.   It does not necessarily follow that, because there is an apparent regularity in the construction of this universe of stars, it is infinite.   It is quite possible for it to be of infinite extent and yet fill only a limited part of space. But as we are entirely unable to make observations through infinite space, we are by no means entitled to infer a veritable creation, an arising out of nothing, merely because of the necessity we feel of assuming a commencement to that universe known to us.

In any case it is highly probable that our universe has, even in comparison with

the duration of the longest star-life, existed
for an exceedingly long time, and that the
above sketched course of evolution of the
celestial bodies has been repeated a
countless number of times. As moreover
the number of stars is very large and their
masses must have been distributed by
chance—and as we may, further, assume
that the evolution of celestial bodies which
consist everywhere of the same substances
and are governed by the same forces takes
place always in a perfectly uniform manner
as has actually been proved by spectral
analysis (in all probability mainly according
to the nebular theory of LAPLACE), we are
able to estimate with tolerable accuracy
the duration of the typical stages from the
observed proportion which exists between
the number of stars of the different classes-
—the spectroscope enabling us to make
these observations, entirely independent of

our subjective sense, of the colours and of the relative luminous power of the stars. Thus, with tolerable accuracy, the duration of the white stage may be fixed at $58.5^0/_0$, that of the yellow stage at $33.5^0/_0$ and that of the red stage at $8^0/_0$ of the mean total luminous existence of a star. The nebular stage must have lasted for so short a time (less than $0.1^0/_0$ of the total time of evolution) that it need not be taken into account, and the same with those stars which exist in a state of transition between different stages, their number being but relatively small.

The great uniformity in the development and in the composition of each class *

---

* A most remarkable proof of this extraordinary uniformity, extending also to the conditions of density and temperature, has been given by J. SCHEINER in his "Die Spectralanalyse der Gestirne," pp. 280—287.

As regards other conceptions relating to the evolution of

justifies us in applying also to our sun the mean proportion as allotted to the different stages of star-life. The sun must have passed the greater part of its existence in the white stage, and, as it is at present in its yellow stage, it must have passed at least 58.5%, or about $^3/_5$, of its star-life. In which term of the yellow stage it is now can neither astronomically nor physically be determined with any accuracy;* but here we may look to palæontology for information.

the stars than that here set forth, which is founded on the ground of the well confirmed nebular theory of KANT and LAPLACE, such as were enunciated by J. NORMAN LOCKYER (The Meteoritic Hypothesis. London 1890) and also by E. W. MAUNDER (Journ. Brit. Astron. Oct. 1891), I refer to SCHEINER (Op. cit., pp. 330—331).

*The spectralanalytic examination of the sun still proves that the sun, like α Aurigæ (Capella), which in all respects resembles the sun and many other yellow stars, must be a very young member of its class and is immediately connected to the white stage by a few transitory stars like Procyon.

During the longest period of the existence of organisms on the earth, as far as our knowledge extends, it must have received considerably more heat than at present. Hence the solar radiation must then have been considerably more powerful, so much more so than at present, that the sun must necessarily have been in quite a different condition. Now, since from the sun's history we learn that, during the greater part of its existence as a white star, it was much hotter, and on passing relatively rapidly from the white to the yellow stage, it lost much of its heat, we see, on the other hand, that the heat received by the earth underwent the same changes, for after a very long period of warmth, a relatively rapid cooling set in, finally reaching the present condition—then we may assume that the period of the cooling down of the climates coincided with the

8

transition from the white to the yellow stage, the period of rapid cooling of the sun. The present thermal condition at the earth's surface was developed during the Tertiary period and was fully attained at the beginning of the Pleistocene period — a date, geologically speaking, so recent, that the time which has since elapsed may perhaps be estimated as but one fiftieth of the time which elapsed since the beginning of the Palæozoic age. We, therefore, may further conclude that our sun only comparatively recently entered its yellow stage, and we may, without committing a great error, consider that it has now passed $3/5$ of its life as a luminous and heating star and that it has still $2/5$ before it.

According to the science of to-day the sun draws nearly all its energy, as HELMHOLTZ first pointed out, from the gradual shrinking

of its gaseous body, ever since the time of
the nebula, from which, according to the
theory of KANT and LAPLACE, it originated.*
HELMHOLTZ and Lord KELVIN† calculated
the time during which the sun can have
been shining with its present intensity,
and how long it can still continue to
do so. HELMHOLTZ estimated the past
period of solar heat at 20 million years.
His calculation was based on the older
value of the intensity of solar radiation
determined by POUILLET (Solar con-

*H. VON HELMHOLTZ, Vorträge und Reden. Vol. 1.
Braunschweig 1884, p. 25: "Ueber die Wechselwirkung der
Naturkräfte," lecture delivered in 1854. Ibid.: Vol. 2, p. 57:
"Ueber die Entstehung des Planetensystems," lecture delivered
in 1871. — The meteorites falling into the sun are to be
considered as quantitatively unessential for the maintenance
of solar energy.

†SIR WILLIAM THOMSON, The Sun's Heat. Proceed. Roy.
Institution of Great Britain, Vol. 12, 1887, p. 1; reprinted
in his: Popular Lectures and Addresses, Vol. 1: Constitution
of Matter, second edition, London, 1891, pp. 376—429.

stant=1.76); according to later investigations by LANGLEY (Solar constant=3)* the sun could for 12 million years only have emitted as much heat and light as at present. LORD KELVIN thus came to the conclusion that, taking into account everything by which the computation could be somewhat increased, "it would be exceedingly rash to assume as probable anything more than 20 million years of the sun's light in the past history of the earth or to reckon on more than 5 or 6 million years of sunlight for time to come." Hence the

---

*S. P. LANGLEY, Researches on Solar Heat and its Absorption by the Earth's atmosphere. A Rapport of the Mount Whitney Expedition. Professional Papers No. 15 of the Signal Service. Washington 1884.—CROVA (Comptes rendus de l'Académie des Sciences. Paris. T. 108. 1889, p. 35) SAWELIEF (ibid, p. 287; T. 110, p. 235, T. 112, pp. 481 and 1200) and R. ANGSTRÖM (Wochenschrift für Astronomie und Meteorologie. 1890, Nos. 14—16) concluded still higher values for the Solar Constant than LANGLEY.

sun is at present in such a state of contraction that it has spent already from
$3\frac{1}{3}$ to 4 times as much heat as it has
left.

The sun's yearly loss of heat must,
however, have been much larger during its
white stage than in its present condition.
It can, therefore, only have been shining
for a much shorter period than 20 million years. Distributing a quantity of
heat from $3\frac{1}{3}$ a to 4 a over 58.5 time
parts of the white stage, and the quantity
of heat 1 a of the collective yellow and
red stages over 41.5 time parts, we find
that during the white stage the sun must
have developed on an average a 2.36 to
2.84-fold intensity of radiation as that
which it has available for the yellow and
red stages. We thus further find for the
white stage of our sun 7 to $8\frac{1}{2}$ million
years,—assuming that the sun will keep

shining with its present energy up to the end of its luminous existence. In reality, however, the solar radiation will be considerably less intense during its red stage, that is, during one fifth part of its future existence as a luminous star. Assuming for the intensity of radiation in the red stage the smallest theoretically possible value, namely zero, then we obtain for the white stage, as a minimum value, from 1.91 to 2.29 times its amount in the yellow stage, which means for the white stage a period of $8\frac{3}{4}$ to $10\frac{1}{2}$ million years. But in reality the intensity of radiation of the red stars is by no means equal to zero.*

*Judging by their spectrum the sun-spots must be in a condition very like that of the red stars (SCHEINER, Die Spectralanalyse der Gestirne, p. 315). Now LANGLEY has found that they emit about 54% of the heat emitted by an equally large part of the photosphere (C. A. YOUNG, The Sun). As, however, the rays emanating from the red stars

We have, further, every reason to suppose that, just as in the white stage (judging from the existence of a numerously represented white class as well as from the general stability of the Pretertiary climates), so also during the yellow stage the radiation of the sun will diminish but very slowly. When LORD KELVIN calculated the future amount of sun heat at 5 or 6 million years, he had, moreover, already taken into account the possibility of a reduced radiation.   Hence the past period of solar heat cannot have lasted much longer, and

and from the sun spots undergo slightly less absorption in our atmosphere than those emanating from the yellow stars and the photosphere of our sun, we may conclude that the red stars possess about half the intensity of radiation of the yellow stars. By this assumption we find a radiation in the white stage, from two to two and a half times that in the yellow stage, and we further find for the first stage of our sun a period of 8 to 10 million years, for the yellow stage of 4 ¹/₂ to 5 ¹/₂ million and for the red stage of 1 to 1 ¹/₂ million years.

we cannot be far wrong in assuming for
the past a maximum duration of ten
million years, and a radiation in the
white stage twice as intense as the present. *
This excess of solar heat must certainly

* Repeated attempts have been made to arrive at some
conclusion about the relative intensity of radiation of stars belonging to the first and second spectral classes, from the photometric
data concerning stars of which the distance and the mass
could be approximately calculated and which could thus be
compared to our sun. The unreliability of these data is,
however, so great that it is quite impossible to arrive at
any trustworthy results from them, even for stars like α
Centauri, for which the data differ the least. (Compare the
Tables of fixed-star parallaxes by J. A. C. OUDEMANS in:
Astron. Nachr. 1890, Nos. 2915—2916; more unreliable
still are the computations about the mass of the double
stars). Hence the great differences in results. Recently E.
W. MAUNDER (Journ. Brit. Astron. Soc. Oct. 1891) found,
in comparing the luminous power of nine stars of the first
class and thirteen stars of the second class, whose parallaxes
were determined by heliometer measurements by ELKIN, that
the luminous power of the stars of the first class is in the
mean only equal to ²/₃ of that of the second class. But
probably the white stars have still greater relative luminous

power, and to this circumstance it must certainly be ascribed that the total luminous power of the white stars of first and second magnitude exceeds somewhat more that of the yellow stars of the same magnitudes than is the case with the stars of smaller magnitude, taking into account the arithmetical proportion of the two spectral classes.

Then, again, photometrical comparisons do not give much information about the relative intensities of radiation; for not only are the rays emitted by the bluish white stars, absorbed in our atmosphere in quite a different manner from those of the yellow stars, but also our eye is not as sensitive to blue as to yellow rays, and is almost entirely blind to a great part of the rays emitted by bluish stars. More sensitive to, and thus a better measure for the real radiation of the white stars is the photographic plate; the maximum of the intensity of radiation of the spectrum of these stars must nearly coincide with the photographic maximum in the spectrum. It appears, however, that the difference between the photographic and the optical brightness of the white stars is but very small—smaller than 0.1 magnitude according to KAPTEYN (Proceedings of the R. Acad. of Sciences at Amsterdam from April 2, 1892)—which is to be accounted for by the fact that more than half of the photographic rays are absorbed in our atmosphere. Of the rays of the yellow stars, however, but one fourth is absorbed. Moreover the eye is at least twice as sensitive to yellow as to blue rays.

HUGGINS and STONE tried (about 20 years ago) to measure

have been sufficient to exercise a noticeable
effect on the general thermal conditions of
the earth.

directly the radiating heat of the stars. These first attempts
made with the aid of the thermopile and the galvanometer
were, however, lately repeated with a much more sensitive
instrument of a different construction, the radiomicrometer, by C.
V. Boys (Proceedings of the Royal Society. Vol. 47. London
1891, p. 480), with the result that, notwithstanding the
extraordinarily high sensitiveness of the instrument, no trace
of an existing star-radiation could be detected.

It, however, can be deduced, with perfect certainty from
all our present data, that the specific radiating capacity of
stars of different spectral classes cannot differ considerably,
and that the white stars can be, and very probably are, at
most, only a few times hotter than the yellow stars and
these again than the red stars. This surprising conformity
with the radiated heat arrived at in the present essay in
quite a different manner (the computation might indeed have
given for result 10, 100 or 1000 times higher values), proves
as well the correctness of the HELMHOLTZ contraction theory
of the sun's heat, as also of the assumption that the sun
only recently, geologically speaking, viz. at the end of the
Tertiary period, became a yellow star, and this seems to me
to be a strong argument that the explanation of the geological
climatic changes given here is based on a sure foundation.

Now, the atmosphere possesses, in its variable amount of aqueous vapour and in its changeable cloudiness, powerful means of regulating the quantity of heat arriving at the surface of the earth, and also its diffusion over the earth in time and space. At the time when the sun still belonged to the stars of the first spectral class,[*] the earth must have probably been in a similar condition to that in which, according to the astronomers and the spectralanalysts, our sister-planet Venus now is, which receiving about twice as much radiated heat as the earth, has a much denser atmosphere, very rich in aqueous vapour and in its higher parts filled with a thick layer of clouds. The thick layer

---

[*] Only the *albedo* of the earth's atmosphere must, because of the greater coefficient of absorption of the sun's rays then, have been much smaller than the present *albedo* of Venus, which is determined by ZÖLLNER and SEIDEL at 0.6.

of clouds must then have reflected a great part of the sun's rays and at the same time this must, together with the aqueous vapour contained in the atmosphere, have made the temperatures more equal in time and space.

That part of the surplus of radiated heat, absorbed by the atmosphere, will almost entirely have benefited the higher latitudes, owing to the stronger general circulation of the air which it must have caused. Moreover, it is of great significance that the sun in its white stage possessed not only a much intenser radiation, but also in the form of rays of a reduced wave length. For these, according to the investigations of LANGLEY, are absorbed in a much higher degree by the earth's atmosphere than the rays of greater wave length. Thus from the white sun, so much richer in violet and ultra-violet rays, far fewer rays will have reached the earth's surface than from the present sun, being in the condition of a yellow star and possessing relatively many red and ultra-red rays. Hence a relatively much smaller part of the energy of the white sun could by direct radiation influence the temperature of the climate anywhere, but especially near the equator. On the contrary, the energy of the blue, violet and ultra-violet rays, which were then preponderating, must, by continuous absorption, have been transferred to the higher atmospheric strata and used in increasing the general circulation of the atmosphere and thus, in an indirect way, have benefited the subpolar latitudes.

If, on the contrary, the sun were to be in the less warm condition of a red star, a relatively small part of its energy of radiation would be spent in maintaining

the general air circulation. The higher
latitudes would then have been indirectly
much less warmed than now, although,
near the equator also, the temperature
would be lower, the increase of red and
ultra-red rays being smaller than the
diminution of the violet and ultra-violet
rays.

The comparatively small number of stars
which, like Procyon, form the transition
from the first to the second stage, must
correspond to the relatively rapid gradual
cooling of the climates of the Tertiary
period. But at the beginning of the Pleistocene period, when the reign of the
yellow sun had commenced and the present
climatic conditions had already set in, twice
or thrice a considerable temporary cooling
took place, leading to glaciation everywhere,
where favourable conditions existed. These
Pleistocene coolings can not have been

much less than the rise of the temperature was during the Pretertiary ages, and must correspond to very considerable diminutions of the solar radiation, which again, being accompanied by a great disturbance in the conditions of the sun, would have become apparent by its colour and its spectrum.

What may then be the stars, at present in this glacial intermediate period of the second stage? Apparently no other than transition-stars from the second to the third class, or stars belonging to the third class itself.* The actual occurrence of

---

* More probably transition-stars,—firstly, because the difference between the temperature of the present climate at temperate latitudes, and that obtaining during the Glacial epoch, is smaller than that of the middle Tertiary (the former being only about the half of the latter), while the red stars are nearly as much cooler than the yellow stars, as the white stars are warmer than the latter; secondly, because during the Pleistocene Glacial epoch the difference

such considerable periodical changes in
the sun's radiation is rendered probable
by the relatively great number of transition-stars between the second and the
third class, by the fact that a number
of yellow stars undergo periodical changes
of shorter duration, by which the breadth
and the obscurity of the lines increase and
even traces of column-shaped zones appear,
like those characteristic of stars belonging
to the third class. Further, by. the fact

of temperature with the present was only about four times
as great as the oscillation of the temperature during the
35 yearly period discovered by BRÜCKNER, of which no
corresponding changes in the sun have as yet been discovered,
though there are many indications that this is also caused
by variations in the solar radiation; thirdly, because the earth
is at present in a period somewhat cooler than the middle
Interglacial periods and corresponding climatically more with
the transition periods between the Glacial and Interglacial
periods, (BRÜCKNER, Klimaschwankungen, p. 314), while the
sun represents at present the pure type of a yellow star.

that many variable stars belong to the transition group and most of them to the third class, and lastly in the body of our sun short periodical changes take place, of which at least an eleven-year period of sun-spots is proved to a certainty. When such periodical changes of short duration take place, periods of longer duration become probable, with which not only the statistical data regarding the different classes of stars are in accord, but also our knowledge of the physical and chemical conditions of the sun, as well as the geological phenomena, give indications. Since the sun entered its present, already considerably cooler state of a yellow star, thereby causing the appearance in its atmosphere of a great number of chemical elements, conditions favourable for periodical changes in its radiation were certainly created by the possibility of chemical combinations being formed. In the last years it has been repeatedly pointed out that the variability of the eleven-year period of sun-spots may also be chemically explained. The geological phenomena of the Pleistocene period can in this way be well explained. Though the estimates for the duration of the whole period have led to very different results, it may be taken for granted that the Interglacial period must have been of very long duration, every one much longer than the Postglacial period in which we are now living. This may, therefore, quite well, even as far as regards its duration, be looked upon as the equivalent of the Interglacial periods with which it entirely agrees in its climatic conditions. It is likewise probable that the Interglacial periods have continued absolutely longer than the Glacial epochs, and that thus their total duration—past and future—may be compared to the number of yellow stars, and that of the collective Glacial epochs to the number of the transition stars to the red (perhaps, but improbably, the red stars themselves included). Everything seems to indicate that during the yellow stage in long oscillations, always during a relatively short time, chemical combinations occur by which the colour of the star becomes reddish (or red), and broader and darker lines, indicative of a denser atmosphere, and dark bands or columns, indicative of chemical combinations, appear in the spectrum. By the otherwise unimportant and regular decrease of the radiation during the yellow stage, these oscillations will repeat themselves probably for a long time without any considerable lengthening or shortening of the intermediate periods, and only shortly before the end of the sun's life the intermittent cool period will grow rapidly and the body of the sun grow permanently red and at last dark.

The Solar Heat arises most probably from Contraction. Seeming Inconsistency with Geological and Biological Data.

According to HELMHOLTZ and LORD KELVIN we must look for the origin of solar heat almost exclusively to the potential energy which the sun possesses by the mutual attraction of its component parts. Before the matter of which the sun consists coalesced and became hot, it must have existed in space in the form of cold masses. According to HELMHOLTZ and Lord KELVIN the force which caused them to coalesce lay in the gravitation or general mutual attraction of the celestial bodies, and on this supposition they based their calculations of the duration of the sun's heat. Lord KELVIN * enquires whether other forces might not have caused the coalescence of the sun's mass. There might have been, so says the great physicist, two cold bodies which came into contact by the velocity caused by their mutual attraction. "With enormously less of probability" there might have been two masses which collided with considerably greater velocities than the velocities due to mutual gravitation. In this case their own motions ought to have been almost perfectly directed towards each other's centre of inertia, for, even with relatively small velocities, a slight deviation from the line joining the centre would prevent the collision or cause the already collided bodies to separate again and make

---

* Sir WILLIAM THOMSON, The Sun's Heat, l.c. p. 13 sqq.

them revolve round the common centre of inertia.

" The dry probability of collision between two neighbours of a vast number of mutually attracting bodies widely scattered through space is much greater if the bodies be all given at rest, than if they be given moving in any random directions and with any velocities considerable in comparison with the velocities which they would acquire in falling from rest into collision." In the latter case the probability of collision will be smaller the greater the velocities of their own motions are. In reality many stars and also our sun have a motion of their own, whose velocity is, however, generally very small compared to that which a body would acquire by the attraction of the sun. While the latter is 612 K.M. per second, there appear in the firmament only few motions

of greater velocities than the tenth part
of this, and that of most stars is much
smaller. * Only a very few stars, as
1830 GROOMBRIDGE, 9352 LACAILLE, 31
Cygni and a few others have so rapid a
motion that it seems this cannot have
been caused by gravitation alone, at least,
taking, with NEWCOMB, † the visible stars
only into account.   But in reality the total
mass and number of the dark bodies in
space is probably much greater than
that of the luminous bodies: the stars, as

---

*According to photographic spectral observations by H. C.
VOGEL at Potsdam, and also accurate direct measurements
of displacements of spectra by J. KEELER of the Lick-
observatory at Mount-Hamilton (Publications of the Astronomical Society of the Pacific, II, p. 265), it is apparent that
the velocity of the motions of a great number of stars is
much smaller than had been determined formerly, for instance
at Greenwich. On an average it was found to be only the
half of that, namely 17 K. M. according to the determinations
at Potsdam.

†NEWCOMB, Popular Astronomy, 1878, p. 487.

these can only be formed by a rare coincidence of favourable conditions. It may therefore be that even 1830 GROOMBRIDGE owes its velocity to gravitation alone. However, relatively considerable velocities are of very rare occurrence in the firmament, and therefore it becomes still more improbable that these could in an important measure account for the energy of the sun and of the other stars.

LORD KELVIN supposes further, that our sun may have originated from a great many bodies moving in different directions and even then the probability of collision must become smaller the greater the resultant moment is of these velocities. In that case also the most favourable condition for collision is that of rest and the most unfavourable that of great velocities of proper motion.

In opposition to this gravitation hypothesis, which is based on the theory of probabilities, J. CROLL poses his impact hypothesis.* He believes that the bodies of which the sun is composed had a motion of their own, much greater than that which could possibly have been caused by gravitation alone, and—though acknowledging that a collision would then be much more improbable—he perceives in this very improbability the proof of the correctness of his assumption. According to his opinion collisions must be very rare, or else the universe could not exist in the condition which we actually observe. It appears to me that the grounds on which he bases his assumption are extremely weak, for the

---

*See: Climate and Time in their Geological Relations. London 1875, p. 353 sqq.; Discussions on Climate and Cosmology. Edinburgh 1886, Chapters XVIII and XIX and: Stellar Evolution and its Relations to Geological Time. London 1889.

probability of a collision of celestial bodies having the velocities he supposes becomes so exceedingly small, that, notwithstanding the great number of dark bodies in the universe and the long duration of individual star-life, scarcely any star could be seen. CROLL has been led to this "impact-theory" by his hypothesis for explaining the Pleistocene glaciation. This hypothesis requires a longer time for the existence of living beings upon the earth than the solar heat can possibly last, supposing this to have originated from the generally accepted source. The chief proof for the correctness of his views CROLL finds in this, that geologists and biologists require a longer time for the formation of the geological strata and for the development of life upon the earth than the 20 million years which the physicists could grant as maximum. By measuring the sediments deposited by the rivers we can estimate the average denudation of the continents, and thus, supposing the present conditions to have prevailed in the past also, we can compute the time required for the deposition of those sedimentary strata of which the mountains of the earth are now composed. These estimates are, however, but very approximate and must lead to very different results. It is, further, possible and even very probable that in former ages denudation took place to a far greater extent than now. It is dependent in the first place on the quantity of rain and consequently on the dampness of the air. Denudation and the disintegration of the rocks, which accompanies it, are undoubtedly most considerable in the tropics, and we actually notice that the rainfall greatly increases towards the equator; according to MURRAY, it is from two to three times as heavy in the tropics as in temperate latitudes. * At the time of the white sun radiating so much more heat, the amount of vapour contained in the atmosphere must have been exceedingly great, and over the whole earth, for when the regions in high latitudes were, compared with the tropics, still relatively short of heat directly received, this was compensated by very warm ocean-currents and mild winds, the vapour of which must have condensed abundantly on the land. The weathering processes are, moreover (the moistening relatively unaltered), really promoted by a

*J. MURRAY, On the total annual Rainfall on the Land of the Globe. Scottish Geographical Magazine. Vol. 3. 1887, p. 65.

In the Malay Archipelago the rainfall is, for an equal area, even five times, at the equator, at an average, four times as large as it is in Europe. Moreover, the rainwater in the tropics contains ten times as much nitric acid as in Europe.

higher temperature, and, besides, the vegetation is an important factor.* During the time when the sun still poured forth an intenser flood of heat on the earth, the circulation of the water must consequently have been more powerful, and therefore its mechanical effects of erosion and of transport of the dissolved and solid products of weathering more considerable. We shall, therefore, certainly not be exaggerating in assuming that at that time denudation may have been ten times as strong as it is now in temperate lati-

---

*F. VON RICHTHOFEN in: G. NEUMAYER'S Anleitung zu wissenschaftlichen Beobachtungen auf Reisen. 2. Auflage. Vol. I Berlin 1888, pp. 230—231.—Compare also: J. C. RUSSELL, Subaerial Decay of Rocks. Bulletin of the U. S. Geol. Survey. No. 52. Washington, 1889.

After MÜNTZ has shown that Bacteria (*Nitromonas*) are the chief agents in the weathering of the rocks, this influence of the moisture and the temperature has become quite intelligible.

tudes—and hence the computations of geological time are not so reliable as CROLL and others suppose them to be.

On biological grounds we may relatively more readily accept a very long duration of the solar heat. The evolution of the organic world progresses extremely slowly. The development of the three-toed *Hyracotherium (Orohippus)* into the true horse, which can be well traced, required almost the whole of the Tertiary period. Bats and whales, these strange modifications of the mammal type, are already found completely developed in the Eocene formation, and at this rate the long time from the beginning of the Palæozoic epoch would certainly not have been sufficient to account for the development of the whole class of mammals. The *Multituberculata*, geologically the oldest known mammals, already existed in the

Triassic period, although probably related and perhaps belonging to the lowest organised branch: the *Monotremata*, which still stand morphologically so far above the lower organised classes of vertebrata and are from their first appearance so much specialised, that, with WALLACE, we must date the origin of the mammals, at the lowest estimate, far back into the Palæozoic age. * According to MARSH and FÜRBRINGER the same is very probably the case with the birds. †
The commencement of power of flight in the ancestors of this wonderfully dif-

---

* A. R. WALLACE, Island Life.   London 1880, p. 205.—
Compare also the beautiful address on Palæontology and the Doctrine of Evolution by TH. H. HUXLEY.   Anniversary Address to the Geological Society of London in 1870.
Proceed. Geol. Society, 1870.

† O. C. MARSH, Amer. Journ. of Science, (3) Vol. 22, 1881.—MAX FÜRBRINGER, Untersuchungen zur Morphologie und Systematik der Vögel.   Amsterdam, 1888, p. 1571.

ferentiated class certainly occurred in the beginning of the Jurassic, perhaps in the Triassic, period. * We have as yet been able to follow the evolution and detect the origin of smaller groups only and not of the chief types or classes. The more we get acquainted with early life, the farther back in the past we must place the first appearance of the higher forms. As to plants, the Dicotyledons were first supposed to have originated in the Tertiary period, afterwards in the Cretaceous period, and now this high class has been found in the Jurassic formation also.† As a rule the oldest representatives of any class do not represent the beginning of its evolution, and DARWIN considers it indisputable, that before the formation of

* FÜRBRINGER, l.c.

† G. DE SAPORTA in: Revue générale de Botanique, Vol. 2, 1890.

the lowest Cambrian strata long periods elapsed, as long as, or probably for longer than, the whole time which has elapsed since the Cambrian period until now, and that during those immeasurable periods the earth swarmed with living creatures. WALLACE, who perhaps more than any other is entitled to express an opinion in this matter, thinks that the time which has elapsed since the first appearance of life on earth cannot have been much less than twice and a half the time which has elapsed since the Cambrian period, and that, in all pro- bability, it has been even longer, for, as is believed, the reaction of the organisms on changes in the surrounding conditions must have been less intense in low and simple than in high and complex forms of life. Hence the organic evolution may for countless ages have progressed extraordinarily slowly.*

* WALLACE, Island Life, 1880, p. 205.

It may, nevertheless, have been that organic evolution under the reign of the white sun progressed faster than for instance in the Tertiary period, in which we can accurately follow the evolution of some forms, and when the light and heat of the sun were already diminishing. The evolution depends in the last instance on the appearance of numerous and manysided hereditary individual modifications of the type, and the mean form of every species and what was probably the principal favourable condition for these physiological and morphological modifications lies, according to DARWIN, in an excess of nutrition;* yet just during the Tertiary period the conditions of nutrition underwent a continual limitation. The more favourable Pretertiary conditions may have acted in yet another way to promote

* Chapter XXII in: The Variation of Animals and Plants under Domestication.

organic evolution, namely by the increase of the number of individuals of every species available for the selection, while with the number of the individuals the variability itself must have increased in a high degree, according to WEISMANN's heredity theory, in consequence of increased intercrossing. The damp uniform warmth of the atmosphere and the greater intensity of the sunlight must have everywhere to a great extent promoted vegetation, and thereby animal life. Palæontology affords many direct evidences that a richer development of the latter has really existed. At least from the middle of the Tertiary period the organic evolution must, therefore, have progressed considerably slower.*

* Principally towards the end of the Tertiary period, when the sun passed rapidly into its yellow and then into its Glacial stage. Here then is a plausible solution of the great enigma, as NEUMAYR calls the phenomenon, of the extinction

over the whole earth of the rich Tertiary mammal-fauna and
its numerous great Diluvial descendants, owing to which
we now live in the midst of a very impoverished fauna—
"an occurrence which, however, shows us that in some not
yet recognisable manner and by still unknown causes, important
changes in the conditions of life prevailed over the great
majority of continents of the earth during the Diluvial
period, influences of more general signification and extent
than any causes" (such as too far going specialisation, direct
influence of the Ice age or extermination by man) "which
one would be inclined to assume for them". (M. NEUMAYR,
Die Stämme des Thierreiches. Wien und Prag. 1889, p.
149). By diminished variability and limitation of the material
for the selection the continually necessary faculty of adaptation—that without noticeable morphological variations included—to changed external conditions must have decreased
more and more, yet at that time these external conditions
underwent such important changes in consequence of the
diminished solar radiation. In the long run this must necessarily have led to the extinction of many species, in the first
place of those of very large animals, such being poorer in
individuals, less reproductive and requiring a greater quantity
of food. Only in this way does it appear conceivable how
so many species in all parts of the world became comparatively rapidly totally extinct without a struggle with the
victoriously competing organisms, and that not merely a
diminution in the number of individuals of every species has
taken place.

Animal life also may have been directly promoted by the intense light of the white sun. MOLESCHOTT, PFLÜGER, VON PLATEN and others had formerly found that light promotes the vital combustion in animal organisms and tissues. MOLESCHOTT and FUBINI had, further, observed that even with blind animals blue-violet light is quite as active as white light, while red light has much less effect.* Still later FUBINI and SPALITTA† observed that the minimum effect of blue and violet light was in mammals and birds, though in toads these colours had the maximum effect. PFLÜGER had suggested before that this strong increase of vital combustion might be caused indirectly by reflectorily excited muscular motions or by the intensifying of the muscular tonus. The experiments

---

* MOLESCHOTT'S Untersuchungen. Vol. 12, pp. 266—428.

† Ibid. Vol. 13, p. 563.

of LOEB with the pupa of butterflies* have
really proved that the oxidation in the
tissues increases only slightly when muscular
motions and changes of the tonus are
excluded. The intensified vital combustion
must still to a small degree be ascribed
to the direct influence of light, and MOLE-
SCHOTT and FUBINI—as already mentioned—
observed that even blind animals were
affected by light, though in a less degree;
which was later also confirmed by MARTIN
and FRIEDENWALD.† It may, therefore,
be taken for granted that light generally also
directly promotes vital combustion, and
it is almost certain that this action is
mainly due to the more refrangible rays

---

*PFLÜGER'S Archiv. Vol. 42, p. 393.

†H. N. MARTIN and J. FRIEDENWALD in: Studies from
the Biological Laboratory of the John Hopkin's University,
Baltimore, Vol. 4. 1890.

at the violet end of the spectrum, the so-called chemical rays.*

Moreover, the influence of light on the growth of plants is well known. This appears mostly in the hindrance of longitudinal growth (etiolation and positive heliotropism), less often in promoting it (negative heliotropism). This influence is nearly entirely exercised by the rays of greater refrangibility, and its maximum corresponds to the transition from the violet to the ultra-violet. So here again we find a pronounced influence on the vital interchange of matter of light in

---

* E. YUNG (Archive des sciences physiques et naturelles. Genève, 1883, pp. 55—56) observed the development of eggs of fishes and frogs to be mostly promoted by blue and violet light.—For the heliotropism of pelagic animals the stronger refrangible rays of the spectrum proved to be more active than the less refrangible rays. (T. J. GROOM and J. LOEB in: Biologisches Centralblatt. Vol. 10. p. 1890).

general, and principally of the "chemical" rays of the spectrum.

The vital interchange of matter and consequently all the vital processes—one of which is the phylogenetic evolution—may therefore have been more marked in animals and plants under the white sun.

In inorganic nature the substances which are sensitive to light are decomposed almost exclusively by the more refrangible rays of the spectrum, chiefly the blue, the indigo, the violet and ultra-violet rays. Now it is very remarkable that, on the contrary, the decomposition processes of the principal substances, or by means of these substances which are sensitive to light in organic nature—the "Sehpurpur" in the retina of the eye and the chlorophyll in the green parts of plants which establish such an intimate relation between light and life, decidedly the most intimate between any phenomenon of the surrounding world and its organism, are not chiefly affected by the " chemical " rays at the violet end of the sun-spectrum, but by the most energetic rays. The curve representing the value of brightness of the colours of the spectrum attains its zenith near the middle of the spectrum and falls towards both ends—and the curve representing the process of decomposition of carbonic acid and water in their elements, the so-called assimilation, that main nutrition-process of the plants which takes place through the agency of chlorophyll, coincides almost completely with that light curve. According to the later determinations made by LANGLEY, by means of the deflection-spectrum and the bolometer,* the maximum

---

*S. P. LANGLEY, Distribution de l'énergie dans le spectre solaire-normal. Comptes rendus de l'Académie des Sciences. T. 92. Paris, 1873, p. 701.

of energy likewise does not lie, as was formerly believed, in the ultra-red but in the yellow, as does the intensity of light in the assimilation of the green parts of plants* I cannot abstain from mentioning this striking coincidence, which might least be expected, considering the character of the material processes. It is one of the most beautiful instances of the adaptation of the organism to its surroundings and conditions under which it lives. Under a yellow sun it must certainly be of great advantage in the struggle for existence if yellow is the predominant colour and if yellow rays promote

---

*HANN (Handbuch des Klimatologie, p. 27) already drew attention to this agreement of the three maxima.

The curve of the energy of the sun's spectrum is less steep, left and right of the maximum, than the two other curves; the sensation of brightness and the assimilation are, therefore, not merely functions of the energy of the rays, but also of their wave-length.

best the nutrition of plants. Accepting this as the cause of the strange phenomenon, we must likewise consider it very probable, that, at the time when the greatest energy of sunlight belonged to blue and violet rays, these, above all others, must have affected the eye and have caused the strongest assimilation in the vegetable world.

Therefore, as at that time the true chemical rays exercised the greatest influence, those most important processes of life, the vision of animals and the assimilation of plants, could be stronger. Thus, the Pretertiary sun may have accelerated the evolution of the organic world in this way also, that by the quality of its light it fitted the animals for a harder struggle for existence, and produced a more luxuriant vegetation.

Further Circumstances in favour of the View that the geological Changes of Climate are to be explained by the History of the Sun's Evolution.

Assuming that the sun at a period, geologically speaking, not so remote, was in the condition of a white star, we may look upon and account for the frequent redblindness as an atavism, a negative inheritance from that long time when the eye of our ancestors was not yet sensitive to red rays which were almost entirely wanting in the white stage of the sun.—The YOUNG-HELMHOLTZ theory of colour sensation agrees with the assumption that the maximum of energy in the solar spectrum has been successively, during a long time, in the neighbourhood of the blue-violet (as late as in the Tertiary period); of the middle of the present spectrum (during

Pre- and Interglacial periods); and of the red (during the Glacial epochs). This theory obtains thereby a genetic and causal foundation. The adaptation for the perception of brightness, acquired during each previous condition of sunlight, was successively preserved, being then useful for the distinction of the quality of the light; the originally monochromatic eye became dichromatic, and this afterwards the present trichromatic eye. The colour-perception was developed as an indirect result of the changes in the kind of sunlight.* Thus the energy-curves of these different solar spectra correspond completely with those curves which represent the intensity of excitation by the three primary colours. Hence it is also clear why red is the least exciting colour,

---

*Consequently we must choose greenish yellow for the middle primary colour.

why in the dichromatic eye the perception of red is most frequently wanting and why the peripheral extension of
the field of vision is smallest for red and
greatest for blue, while, on the contrary,
in the middle of the retina (because of the
yellow colouring around the fovea centralis)
all blue and violet becomes somewhat
obscurer.*

According to experiments by PAUL BERT
and SIR JOHN LUBBOCK, with Daphnidæ,
it appears that the eye of even very low
animals behaves, on the whole, with regard
to colours, in the same manner as the
human eye.†

---

*More than GOETHE himself presumed is there truth in
the saying :

> " Wär' nicht das Auge sonnenhaft,
> " Die Sonne könnt' es nie erblicken."
> If in the eye were not a sun, -
> It never could behold the sun.

†RAPHAEL DUBOIS noted a similar behaviour in the

A beautiful instance of an indirect adaptation of the organism to the present kind of sunlight is given by the phosphorescence of the fire-fly. The strong light emitted by this animal consists nearly entirely of the green and yellow rays of the middle of the solar spectrum; it contains extremely little blue and orange and no red or violet.* That the maximum of energy in this spectrum is situated a little beyond the maximum of energy in the solar spectrum, towards the violet, in the same way as that of the sensitiveness of the eye for rays of different wave-lengths, may possibly be looked upon as an inheritance from the time when the greatest

siphons of *Pholas dactylus*, which, though eyeless, are very sensitive to light. (Comptes rendus de l'Académie des Sciences. Paris, 1889. T. 109. pp. 233 sqq. and 320 sqq).

*S. P. LANGLEY and F. W. VERY, On the cheapest form of light. Amer. Journ. of Science. 1890. (3) Vol 40, pp. 97—113.

energy of the solar spectrum was due to rays of shorter wave-lengths than at present.

We find the views set forth here, confirmed to the widest extent in the vegetable world.

In a similar manner to the colour-perception of the eye is, it seems to me, to be explained the predilection of Nature for green in the vegetable kingdom; namely, as a consequence of adaptation and inheritance. The spectrum of the colouring matter of chlorophyll shows strong absorption in the blue and violet, in the orange and red, less strong in the yellow, very feeble in the green. Therefore this colouring matter in the plants appears green to us. The energy of the rays absorbed by this "sensibilisatory" colouring matter* is changed into energy of

* Compare: H. W. VOGEL, Handbuch der Photographie. Part I. 4th Edition, Berlin, 1890, p. 202 sqq.

a different form, namely not only into chemical action, but partially into fluorescence and heat, as is proved by the fact that the action exercised by blue rays in the process of assimilation is but relatively small, while that of yellow rays on the contrary is the strongest. Though the action of the rays of different wavelengths, manifesting itself in the chemical change of the molecules, corresponds almost entirely to the energy of these rays, so that the assimilation and the energy curves, except in their steepness, differ only in that the curve of the assimilation is shifted a little towards the red end of the spectrum, yet the blue and violet, the orange and red rays exercise in addition a considerably different action on the molecules. In the same manner as the molecular structure of the present chromophyll, by whose agency the

assimilation takes place, has adapted itself
to the light of a yellow sun, so—supposing the views advanced in this essay to
be correct—it must also have once adapted
itself to the light of a sun which was rich
in blue and violet rays, and a sun with
a preponderance of orange and red rays, and
it is natural to consider the strong extinction of just these rays among the
present sun-rays as a reminiscence and
their present modified actions as physiological homologies of the assimilatory processes of those times. The assimilatory
action must always be accompanied by
extinction,* and thus we can imagine how,

---

*According to the DRAPER-VOGEL "photochemical law of
absorption," that only the rays absorbed by a substance, or
by an "optical sensibilisator" can chemically act upon that
substance. (H. W. VOGEL, Handbuch der Photographie.
Theil 1. 4th Edition, Berlin, 1890, p. 58, and J. M. EDER,
Die chemischen Wirkungen des Lichtes. Ausführliches Handbuch der Photographie. 2nd Ed. Vol. 2. Halle 1891, pp.

by a slight modification of the molecular structure, instead of chemical action, work of a different form is effected, and how, when the former decreased, the extinction of the corresponding rays remained, though now mostly caused by fluorescence and heat. In the orange that process of modification seems still to be progressing, and again the relatively strong assimilating action of orange rays is to be considered as a physiological rudiment which remained on the other side to that of the colour-perception, probably because here the power of adaptation, and in the eye heredity chiefly prevailed. During the immeasurably long ages when the specific chemical blue rays were predominant the adaptation to these must have become

153—154).—TH. W. ENGELMANN (Farbe und Assimilation, Botanische Zeitung. Vol. 41. 1883 Nos. 1 and 2) has shown its special validity in the process of assimilation.

very perfect and close, and their assimilating action very strong, and during the time of transition to the yellow stage of the sun, when the blue rays were still very abundant, this adaptation must have lasted long, in a similar manner as at present the adaptation to the orange rays. But because of its great perfection and the strong action of the relative rays, adaptation to the then predominant rays between blue and yellow would have been useless, and this has never, therefore, taken place in a noticeable degree.

Hence the present world owes the beautiful emerald colour of Flora's garment to these great changes of solar radiation.*

* Likewise the red colour of the young leaves of many plants has probably to be looked upon as the transitory and imperfectly inherited condition from Preglacial time. At present the red colouring matter takes no part in the making of fecula; but, according to chemical experiments, still in the making of sugar.

The connection of these with a great morphological change of the vegetable world seems to me to be quite certain. The chief plant types of the Palæozoic age—the Calamites, Lepidodendra, Sigillariae—as well as the Equisetaceæ, tree ferns, Cycads and Conifers of the Mesozoic period, were all sparingly branched and the total superficial extent of their leaves was but small, as is likewise the case with all existing lower (that is old typical) chlorophyll-plants. Among the Angiosperms the Casuarinae (*Chalazogamae*), which according to TREUB's important investigations,* are equivalent to the Mono- and Dicotyledons together (*Porogamae*), and, therefore, are originating from early

* M. TREUB, Sur les Casuarinées et leur place dans le système naturel. Annales du Jardin Botanique de Buitenzorg. Vol. 10. Batavia 1891, pp. 145—231, with 21 Plates.

Mesozoic time, have but very poorly developed organs of assimilation. Although during the Cretaceous period, and even during the Jurassic period, plants existed more richly branched and with a larger total area of leaf-surface (Dicotyledons), these formed only a very inconsiderable part of the vegetation of the earth. It was not before the Tertiary period that, starting from the north pole, their centre of development, they drove entirely into the background the other plant classes. The small extent of leaf surface of the Pretertiary chlorophyll plants can only be explained by the relatively stronger assimilation, since the supposition of a formerly greater amount of carbonic acid in the atmosphere can now no longer be seriously discussed; increased transpiration certainly did not exist in the damp climates of these periods, and simple adaptation of organs so very important to life would certainly have been brought about in the vegetable mantle of the earth long before the Tertiary period, had it been necessary.

Thus, as well in the past of the vegetable world as in that of the animal world,* the majestic effect of that limitation of the solar energy bestowed on our planet becomes apparent, and we see this very sparingness of the sun become the cause of the evolution of the most perfect and highest forms of life. But the whole history of its evolution has been stamped by the brilliant orb of day in every green leaf and in the sense of sight of animals.

* Compare especially pp. 55—59.

THE END.

www.ingramcontent.com/pod-product-compliance
Lightning Source LLC
Chambersburg PA
CBHW021806190326
41518CB00007B/471